Kamaldin Zhumabayev

Improving the management of national natural resources

Kamaldin Zhumabayev

Improving the management of national natural resources

ScienciaScripts

Imprint

Cover image: www.ingimage.com

This book is a translation from the original published under ISBN 978-3-659-69986-3.

Publisher:
Sciencia Scripts
is a trademark of
Dodo Books Indian Ocean Ltd. and OmniScriptum S.R.L publishing group

120 High Road, East Finchley, London, N2 9ED, United Kingdom
Str. Armeneasca 28/1, office 1, Chisinau MD-2012, Republic of Moldova, Europe
Printed at: see last page
ISBN: 978-620-8-03985-1

Brief information about the author

Kamaldin Zhumabaev

Scientific Degree - Candidate of Economic Sciences, Professor;
University - Osh Technological University;
Scientific Interests - Safety in emergency situations. Ecology.

TABLE OF CONTENTS:

CHAPTER 1	**6**
CHAPTER 2	**27**
CHAPTER 3	**56**

Introduction

Kyrgyzstan is in the process of transition from a system of centralised planning to a market economy. However, the inefficiency of the state economy and bureaucratic management methods hinder the normal functioning of the market and do not allow for growth in the welfare of the population. Citizens of the Kyrgyz Republic under the system of planned economy had guaranteed employment and other social guarantees , but the price of these achievements was low labour productivity, falling living standards and, no less importantly, sharp deterioration of the environment as a result of irrational use of natural resources, distorted price system and obsolete technologies.

The long-term goal of the transition process, for the sake of which economic reforms are being implemented in Kyrgyzstan: to create a prosperous market economy, providing a sustainable increase in the living standards of the population. The transition process implies a change in the system itself - reforms should affect the most fundamental laws of society, transform the institutions that determine and guide its behaviour. In other words, we are talking about profound transformations, both economic, social and environmental.

The successes of the planned economy were not insignificant. In reality, however, the welfare state was much less stable because the inherent low efficiency of the centralised planning system was insurmountable. Planners did not have all the information that prices carry in a market economy. The success of planning became dependent on personal relationships and connections. The consequences of this for the environment were most serious, catastrophic.

Transition to the market is a complex process of creating the new, eradicating the outmoded and adapting the existing. Queues have already given way to the market. Scarcity has been replaced by unlimited choice.

Kyrgyzstan was a raw materials appendage during the Soviet period, having received sovereignty, the Republic has gained independence in the free use of natural resources. In this regard, the solution of many environmental problems has become more complicated, as there are no sufficient funds for environmental protection, rational use of natural resources and improvement of management of these processes.

For this to happen, they must be legislated and property rights distributed, and a significant part of the national wealth must cease to be the property of the state.

Central to the transition process are price liberalisation and the expansion of markets, the emergence of new firms and the implementation of programmes designed to establish or maintain stable prices. But this must take into account historical and natural conditions, which have a profound impact on both the relative importance of different market reforms and the ways in which they are implemented.

The key to reform is to secure the potential benefits of resource utilisation by creating a free market. However, under conditions of sharp macroeconomic imbalance and high inflation, free natural resource prices will not be able to become reliable market signals. Therefore, liberalisation of natural resource prices only in combination with stabilisation will be able to ensure higher economic growth both during the transition period and thereafter. This is task one.

The second challenge is to develop property rights and incentive structures and to create a predominantly private economy. The role of natural conditions and resources is also important here.

The third objective, which is important not only economically, socially, but also environmentally, is to reduce poverty and other negative effects of the transition process on certain groups of the population.

Moreover, depending on the natural conditions for rational use of natural resources and the nature of the reforms, the transition process may be accompanied by a reduction in poverty. But at the same time, the situation of many others may worsen as a result of radical changes in the economic system. Addressing the problem of vulnerable populations requires effective social policies and environmental measures that promote sustainable economic growth.

The transition from a planned economy to a market requires a revision of the entire management concept.

In the last few decades, the global interest in the problems of natural resource management has increased dramatically. This is due to the increased role of nature protection and rational use of natural resources in the economic development of countries, material well-being of people and ensuring their health.

In this regard, not only the consequences of violating the laws of nature are widely investigated on a global scale, but also the benefits of environmental management are studied with the necessary certainty. Naturally, man should use natural resources in his own interests. However, from the point of view of economic science, sociology and natural science, many issues of nature management require scientific justification.

Natural resource management is of particular interest for Kyrgyzstan. This is especially necessary for the leading sectors of the national economy - agriculture, mining, non-ferrous metallurgy, hydroelectric power generation.

Due to the collapse of the USSR, the economy of Kyrgyzstan is in a deep crisis, there is a decline in production in all sectors of the national economy. Therefore, the rational use of natural resources to stabilise the economy of the republic is extremely important.

The acuteness of many problems of environment and nature management in the republic and its regions tends to aggravate due to insufficient financial and material resources, changes in the structures of household and industrial consumption, degradation of land and other resources.

At the same time, until recently, the main directions and specific tasks of natural resource management did not touch upon the problems of market relations and freedom of resource disposal in the republic, due to which there were no relevant developments on the formation of economic and legal mechanism of nature management not only at the regional level, but also in the republic as a whole.

The regional peculiarities of natural resources management have not been sufficiently identified, so the issue of developing a general scientific concept of natural resources management, on the basis of which the process of interaction between society and nature could be carried out consciously, purposefully and with optimal results, is considered extremely important.

The essence of natural resources management is to ensure the preservation, restoration and, where necessary, improvement of favourable natural conditions for human life, development of production and culture through scientifically based regulation of society's impact on nature. Here the economic and legal mechanisms of environmental protection should be developed. This issue in the conditions of transition to market relations should be among the most urgent and most important problems of economic and social development.

CHAPTER 1

Methodological bases for improving national natural resources management

1. Content and essence of improving the management of national natural resources

The problems of natural resources utilisation are currently becoming particularly relevant and have become the subject of in-depth research. The fact is that in any socio-economic formation the interaction of society with nature is accompanied, on the one hand, by direct human use of natural matter, on the other hand, by environmental protection measures. Environmental protection measures are especially necessary at present, as many scientists and specialists estimate that environmental problems in the republic have become extremely acute. In particular, over the last 20 years dozens of the most decisive decrees on environmental protection have been adopted. And none of these resolutions has been fulfilled. Not only nature continues to perish. Millions of people are dying and losing their health.

This situation is most often explained by the fact that environmental protection measures are subject to strict budget constraints, there are no reliable and effective projects, there are no competent specialists capable of solving complex environmental problems, etc., i.e. it is most often explained by people's attitude towards nature. Even Russian philosopher N.A.Berdyaev noted the outstanding role of man in environmental protection measures: "The basis of historical progress lies in the relationship of the human spirit to nature and the fate of the human spirit in these interactions with nature.[1]

The problems of nature management have become more relevant since the Kyrgyz Republic gained independence and chose a democratic path of development, as measures to protect the environment, conservation and rational use of natural resources are becoming important in economic development. This, in turn, requires clarification of the content and essence of natural resource management, which, under new conditions, are necessary to ensure economic reforms. It is known that nature in the broad sense is the whole being, the whole world in the diversity of its forms.[2] Nature is an objective reality that exists outside and independent of human consciousness. It has neither beginning nor end, but is infinite in time and space, is in constant motion and change.[3]

Human society in nature is its constituent part. In other words, man is a biological organism of nature, a living creature with thinking, the ability to create and use tools of labour. As a biological factor he is characterised by: metabolism, excitability, growth, development and reproduction. It occupies a special place in nature.

---

[1] Berdyaev N. Entry of the machine, "Nature and Man", No. 8, 1989, p . 64.

[2] Lemeshev M. Economy and ecology : the fatal conflict and ways of its resolution. Voprosy ekonomiki. № 11, 1990 г., 71.

[3] Philosophical Dictionary. Edited by Rosenthal M.M., M., Politizdat, 1975, pp. 329.

If man is placed at the centre of nature, then an environment is formed around him. Or the environment is the environment of human habitation and production activities. It is composed of the natural environment and the sociosphere.

Natural environment is a set of abotic and biotic factors, natural and altered as a result of human society activities, which have an impact on humans and other organisms. It is distinguished from other components of the environment by the property of self-maintenance and self-regulation without corrective human intervention.[4]

Consequently, the state of nature management largely depends on human action, a purposeful process of natural resource management, including by the general public. It should be noted, however, that on the earth the reserves of natural resources are limited. This in turn raises the problem of rational resource use and preservation of the protective properties of nature. Therefore, it is no coincidence that the concept of protecting the place where we live and dwell, i.e. environmental protection, was introduced into the scientific turnover back in the last century. This science is called ecology, which was first introduced by the German biologist Ernst Haeckel in 1868.[5]

In its modern form, ecology covers an extremely wide range of issues and is closely intertwined with a number of related sciences such as biology, geology, astronomy, physics, chemistry, genetics, philosophy, geography, demography, medicine, sociology, law, monitoring, environmental quality management, etc.
Ecology has a profound impact on all aspects of social life, economic and technical progress, and management of public institutions. Therefore, the issues of their financing, forecasting, organisation, efficiency occupy one of the central places of the state policy. In this connection, studies on the organisation of management of the economy of environmental protection have begun to develop. The results find wide application beyond the boundaries of science in solving many economic and social problems.

Environmental problems require solving complex engineering tasks in the field of nature management, protection and control of natural environment pollution. These problems represent a system of social, technical, sanitary and hygienic measures aimed at ensuring safe and healthy living and working conditions for humans.

Another important ecological task is the careful attitude to natural resources, as man in the process of his life activities increasingly extracts masses of natural resources from the natural environment to create more products. At the same time, often economic laws, norms were not respected.

As a result of the negative consequences of such activities, society is now

---

[4] 4-language encyclopaedic dictionary of terms on physical geography. Edited by Spiridonov A.I., Moscow, Sov. encyclopaedia, 1980, p. 300.

[5] Markovich D.J. Social Ecology. M., Prosveshchenie, 1991, p. 5

experiencing a significant shortage in the use of many traditional natural resources. As a result of the negative consequences of such activities, society is now experiencing a significant lack of utilisation of many traditional natural resources.

Today, it is impossible to calculate the damage caused by improper use of natural resources and environmental degradation. Hence the need to measure the use of resources, as well as to find out their manageability. Rationality of resource use for society, with their increasing scarcity should consist in obtaining the largest amount of use values from the mass of raw materials available in circulation in the process of production activities.

The essence of natural resource management is to ensure, through scientifically based regulation of society's impact on nature, the preservation, restoration and, where necessary, improvement of favourable natural conditions for human life, production and cultural development. This issue in the conditions of transition to market relations should be among the most urgent and most important problems of economic and social development. Moreover, under the new conditions, an economic mechanism for regulating nature management should be developed.

The core of the process of nature management is resource use, including the use of mineral resources and products. The use of natural resources should represent a single, to a certain extent acting according to its specific principles and rules, an independent subsystem in the economic activity of society.

One of the most widely promoted and justified nowadays in nature management is the principle of "waste-free" technologies and productions. But despite its effectiveness in many cases, this principle cannot be implemented at hundreds of operating enterprises due to the peculiarities of processes, limited space and organisational and economic shortcomings. The solution of resource and environmental problems lies in the rationalisation of resource use through the construction and implementation of regional integrated programmes, which should be used to modernise existing technologies and equipment, maximum replacement of primary raw materials by waste and the like.

In this connection, a rather specific question arises: Do not the introduction of market relations, not the concentration of production, but, on the contrary, its unbundling, the emergence of more and more small forms of industrial production, small peasants and farmers contradict the solution of the problems of rational use of natural resources?

In our opinion, they are not contradictory. In market conditions, the reconstruction of production, including its unbundling, should be carried out on the basis of progressive technology, installation of new equipment, machines and units in order, on the one hand, to maintain competitiveness, on the other hand, to ensure economic, environmental and social effects. At the same time, where new production is organised, the task of cleaning effluents, gases, waste, harmful emissions, etc. is set. In such a way that a minimum amount of material mass neutral to the natural environment is discharged.

Consequently, the essence of any new and modernised existing production should not be to clean and collect what was released and discarded earlier, but to prevent the formation of such substances that would later require collection and purification. From this perspective, the criterion for selecting new techniques and technologies at all levels should not be the immediate economic effect per se, but the ecological effect, which takes into account all the social consequences of incomplete utilisation of resources. In this regard, first of all, the utilisation of material resources plays an important role. Consequently, in the process of transition to a market economy, raw materials and material resources form the core of nature management, which in turn should be based on the use of best practices and implementation in the practice of production organisation, socio-economic and regulatory requirements, interaction of production with the environment. When developing such projects, the objectives are set to maximise the quality and volume of production at the maximum saving of living and embodied labour, while maintaining the neutrality of technology and the entire production complex to the environment. Such projects of enterprises should be developed for each typical production-technological direction. Further, the parametric assessment of the economic characteristics of the project as a whole and by its elements is carried out. In this regard, the main directions of nature management should be:

- minimisation of material and resource intensity, rationalisation of the structure of material and resource consumption in the production sphere;
- progressive interchange of materials;
- quality organisation of capital and current repairs of machinery and equipment;
- systematic improvement of packaging and preservation of goods, structures, materials, containers, as well as comprehensive protection of metals from corrosion;
- technical and organisational improvement of the system of transport cargoes with acceleration of their mileage, reduction to a minimum of counter transport, losses of mass and marketability of raw materials and resources, etc.

The construction complex plays a significant role in nature management. The lagging capacities, imbalance of the construction complex, as well as a sharp reduction in its work in the post-Soviet period have a negative impact on the rationality of nature management through many channels: primitive extraction of construction materials; non-synchronous development of links of production complexes; underprovision of various communications, including roads and infrastructure facilities, etc.

An important element of natural resource management is the rational and efficient use of resources in the agrosphere. The literature provides various data on the role of agro-industrial production in disturbing the ecological balance. But as studies show, they are approximately the same for four indicators: agro-industrial enterprises pollute agricultural lands by 70%, soils by 60%, water resources by 40%, and atmospheric air by 35-38%. This is due to the fact that the intensification of agricultural production all over the world, including our republic, was based on energy-intensive technical and technological systems. Practice has shown that further growth of energy costs does not give the expected effect. So, if earlier at extensive method of crop production it was possible to receive 20-50 food k/cal per one spent calorie of irreplaceable energy, then at modern energy-intensive technologies - only 2, i.e. 10-25 times less. Consequently, the active use of means and objects of labour per unit area has encountered ecological resistance, a certain ecological barrier. They should be taken into account in the development of fundamentally new technological solutions to increase the production of food resources.

In conditions of wide application in agriculture of fertilisers, herbicides, pesticides, biostimulants, qualitatively new relations in the system "plant-environment" arise, and therefore the degree of risk of disturbance of ecological balance of agrobiocenoses increases significantly.

These and other negative consequences of nature management put a number of tasks on the agenda. One of them is related to the restoration of soil fertility based on the use of organic fertilisers, application of energy-saving and soil-protective technologies of soil cultivation in combination with ecologically sustainable varieties and hybrids of agricultural crops. Ecological measures in farming include not only soil protection, but also water resources protection, i.e. water saving in irrigation networks and under irrigation.

The next task is effective, environmentally friendly plant protection through the use of non-toxic, from rapidly degradable to harmless substances, biological means. Practice shows that the use of organic fertilisers for soil enrichment, promotes humus formation, improves soil mechanical structure, balances physico-chemical and microbiological processes, normalises the absorption capacity of mineral substances, prevents acidity and salinisation of soils, maintains a normal balance of carbon dioxide and has a beneficial effect on the uninterrupted provision of the photosynthesis process of agricultural crops.

Very positive results in terms of ecology are obtained by the introduction of non-waste technologies in processing. Thus, the introduction of production facilities for the production of raw alcohol from grape pomace and yeast sludge from the wine and juice industry, molasses from beet sugar wastes, vitric acid from grape pomace wastes, grape oil from grape seeds, feed flour from meat production wastes, etc., give good results.

Widespread introduction of low-waste and waste-free technologies in the agro-industrial complex not only increases the volume of useful resources, but also directly affects the restoration of ecological balance and increasing the intensity of economic use of natural resources.

All this contributes to clarifying the content and essence of natural resources management and increasing its level on this basis. Of course, the process of nature management has a continuation beyond material production, namely: in the rationalisation of consumption and use of products in the non-production sphere, in the sphere of commodity circulation and service on the basis of scientifically-based standards; in the sphere of wholesale sales, storage of products, retail trade, including, for example, the pricing system with mobile retail prices for agricultural products in the period of its mass harvesting, etc.

It is possible to outline a number of other diverse directions for improving the efficiency of environmental management. However, if we consider three spheres in unity: natural resources, natural conditions, economy, then the main directions of rationalisation of nature use and improvement of management should be considered as follows: comprehensive and full use of primary natural resources with minimisation of technological losses and waste; maximum mobilisation and utilisation of industrial and final consumption waste.

The organisation of measures for environmental protection should be based on the knowledge of the laws of nature and the deeply scientific application of the laws of development in society. On the basis of these factors, society will be able to intelligently manage natural processes and create special mechanisms to direct the development of the biosphere in the direction necessary for human beings.

It is about the content of environmental protection in order to ensure control over the state of the environment and sources of its pollution on the basis of the use of the latest scientific and technical achievements and means, to introduce new effective methods and systems for the development of new mineral deposits, to combat harmful emissions of substances into the atmosphere, industrial, transport and other noises, vibrations, electric and magnetic fields and radiation, to implement measures for the integrated and rational use and protection of the environment, and to protect the environment and the environment.

Another aspect of the problem under consideration is the fact that nature management has a pronounced territorial character, as such important economic and social tasks as rational use of natural resources, raw materials, materials, waste and, finally, maintenance of environmental quality, namely, environmental protection measures, are considered and solved, as a rule, within regions.

2.Place and role of natural resources management in the system of national economy

The process of nature management is a part of the environmental problem, which consists of environmental protection, rational use of natural resources and identification of sources, determination of types and extent of environmental pollution, aimed at maintaining the productivity of health, aesthetic and other dignities of nature in the interests of mankind.

This peculiarity of nature management should be used to determine its place in the system of national economy management. Firstly, by changing nature in the process of transforming existing and creating new artificial ecological systems, man himself undergoes certain changes, but the process of his evolution is very slow. Over the last millennia, the human organism has not practically changed genetically. Human life still requires air, water and food of animal and plant origin, which belong to the biological resources of the Earth. These include the so-called "habitat" - a part of the Earth's surface with conditions favourable to humans.

To meet the social and domestic needs of modern human society, a great variety of substances that are components of the upper layers of the lithosphere, the Earth's mineral resources, are consumed. The production of energy, the consumption of which is extremely high nowadays, is also based on the consumption of the Earth's energy resources.

Thus, the biological resources of the planet determine the possibility of human life on Earth, and mineral and energy resources are the main sources of material production of human society. The shortage of biological resources of the Earth aggravates one of the important problems of ecology - hunger. The economic roots of this problem are lack of food, chronic malnutrition, decreased immunity to disease, lack of protein and vitamins, i.e. all these manifestations arise from food supply problems.

Secondly, man takes from the bowels of nature dozens of times more than necessary to meet his immediate needs, aggravating more and more the shortage of mineral resources. According to N.F. Reimers, if we take into account that all countries will reach the level of the USA in terms of demand, then it will be necessary to increase the production of electricity 6 times, iron 75 times, copper 100 times, lead 200 times, tin 250 times, etc.

Naturally, such an increase is not realistic, as these minerals simply do not exist on Earth in such quantities. This threatens eternal economic inequality of regions, countries and peoples. The shortage of natural resources generates a shortage of raw materials necessary for production, as well as for the economy as a whole. The efficiency of solving social, economic and technical tasks, as well as anthropogenic transformations of nature, is largely determined by energy production and the scale of energy resources. The rates of economic development are in direct dependence on the state of the energy sector. It has a very significant impact on the natural environment, being a source of various types of pollution of water, air, the earth's surface and its subsoil, as well as the main consumer of mineral fuel, determining the level of its extraction.

Hence it is quite clear that in the management of the national economy, first of all, it is necessary to take into account human needs, which can be satisfied through the use of natural resources.

At the same time, in the process of managing the national economy, it is necessary to take into account not only the amount of raw materials and resources extracted from the subsoil, but also the consequences of their further use in human life. A typical example is the energy of the so-called "internal combustion" of liquid fuels, which, for example, propels a vehicle. Energy production leads to the consumption of ever-increasing amounts of fuel and energy resources (FER) and pollution of the biosphere. Energy production associated with the combustion of any type of fuel results in "thermal pollution" of the natural environment. It is known that pollution in ecology is understood as an unfavourable change in the environment, which is entirely or partially the result of human activity, directly or indirectly changing the distribution of incoming energy, radiation level, physical and chemical properties of the environment and the conditions of existence of living organisms. Clearly, such changes can affect humans directly through air and food. They can also affect humans by deteriorating the properties of the things they use, the conditions of rest and work.

Any negative impact on humans through ecosystem destruction, biosphere pollution, etc. requires careful and considered management.

Undoubtedly, the improvement of natural resources management requires the development and implementation of aggregate economic, managerial, legislative measures aimed at a more expedient organization of the production process of production, achieving the highest labour productivity at the lowest cost of money, technical, material resources and natural resources.

Thus, rational, more thrifty nature management simultaneously fulfils two interrelated functions. On the one hand, it satisfies human needs, as natural resources form the material basis for the development of social production and fulfil an economic function. On the other hand, unlike other means of production, which are the embodiment of embodied labour, they form the natural environment and, therefore, perform an exclusively ecological function.[1]

Hence it is clear that the basis of natural resource management should be rational and integrated use, as well as people's actions, which include the restoration and multiplication of natural resources and the creation of protected areas through the withdrawal of certain areas of nature from exploitation, with mandatory compliance with the basic processes of purposeful scientifically based transformation of nature.

Rational nature management in the organisation of production, in the broad sense of the word, gave rise to the concept of "economy of nature management". The economics of nature use covers all aspects of nature consumption, starting from the issues of resource exhaustibility, up to the production of the final product and the use of various kinds of waste and secondary resources.

At the same time, it should be noted that at present there are no clear criteria for the presence of environmental problems in the management of national economy both at the republican level and at the level of primary links. First of all, it concerns environmental protection activities, which are not balanced at the above levels, if only because the mass of waste in production and final consumption is not always taken into account. This is expressed, in particular, in the fact that in real practice the scale and organisation of integrated use of raw materials, involvement of wastes into circulation do not meet the current and, moreover, prospective situation. The level of utilisation of the mass of primary natural resources and wastes of production and final consumption in many sectors of the national economy of the republic lags behind the achieved level in this sphere in some foreign countries.

Economic mechanisms, production management system, farm organisation, methods of accounting, analysis, efficiency assessment, pricing insufficiently guide and stimulate the economic process to form rationalisation of resource use of input resources, extensification, mobilisation and intensification of the use of industrial and final consumption waste.

In the transition period, when market relations are introduced in the republic, often managerial decisions are applied without ecological ideology. Previously existing socialist collective farms - kolkhozes and sovkhozes are being transformed into small-scale farms and peasant farms. Numerous small and medium-sized enterprises engaged in the extraction and processing of natural resources have emerged. This created economic chaos, firstly, by complicating the technology of production processes, organisation of control and preventive measures. Secondly, if we take the agro-industrial complex for example, the technology of storage, processing and canning of agricultural products is not observed. This leads to a decrease in the quality of products. There are frequent cases of poisoning. A certain part of these goods and resources is thrown into waste as a result of spoilage and unsuitability. Thirdly, business plans do not provide for environmental protection measures, and the state nature management authorities are unable to implement them due to the lack of a control mechanism. Fourthly, due to the fact that the production, previously working on imported raw materials, due to the lack of local raw materials and appropriate technology for processing of available natural resources, began to use the available resources using primitive technology, violating environmental standards.

Lack of fuel and energy resources in cities and settlements led to the destruction of ornamental trees, which fulfilled at the same time a residential role.

In the post-Soviet period, the problems of natural resource utilisation in the republic have become more acute, requiring a considerable amount of financial resources. The Kyrgyz economy is experiencing serious difficulties due to the loss of guaranteed markets under the former Soviet Union and budget subsidies from the centre (amounting to about 11% of GDP before independence)[1] .

The use of water resources has always been a problem for the Kyrgyz Republic, which are often at risk, inefficiently allocated, and insufficiently protected from chemical and biological pollutants. Transition problems also include the year-to-year deterioration of land resources. Land degradation (in the form of erosion and soil compaction) and deforestation have long led to a reduction in the yield of all cropped land.

A significant portion of the Kyrgyz economy both before and now depends on the proper management of its irreplaceable mineral resources. The country's mineral resources, which were previously the subject of great expectations in the past (such as mercury, uranium), are now not finding large markets, which exacerbates the problem of financing environmental protection measures from its own resources.

In this regard, in order to improve the management of natural resources in the development of the strategy of economic and social development of the Kyrgyz Republic in the future, it is necessary to put forward requirements to ensure the concentration of resources for the implementation of the most important national programmes, with their rational use, strengthening the regime of economy and elimination of losses in the national economy. In addition, in order to ensure dynamic progress in the structure of the national economy and continuous improvement of the efficiency of all social production, it is necessary to set among the basic conditions the requirements for correct prioritisation of the development of sectors and territories.

It seems that in the field of resource use prioritisation is of exceptional importance. This factor quite noticeably affects the efficiency of functioning not only of the relevant industries, but also of the whole economy. Therefore, in order to improve their efficiency it is necessary to forecast at the enterprise, regional and national levels:

- generation and utilisation of social production waste on a par with primary resources, and for enterprises on a par with the main functions and results of production;

- to recommend to industries and separate productions the consumption of primary resources only after exhausting the possibilities of economically feasible and ecologically necessary utilisation of public production wastes according to the results of the research;

In addition, in the conditions of transition to market relations with the aim of rational nature management, measures should be envisaged to enable environmental design. It is known that the activity of an enterprise begins with a project. According to the project there are coordinate changes in construction, reconstruction, technical re-equipment, development of new products, etc.

At the design stage it is necessary to take into account environmental conditions and regulations that affect technological processes and the state of the environment. Of particular importance are the norms of technological design, through which resource-saving technological processes can be introduced into practice. To the greatest extent it is necessary to develop a regulatory framework in the field of public health protection, in the form of sanitary norms. The list of hygienic norms, maximum permissible concentrations (MPC) of pollutants in the atmosphere, air of the working zone, water bodies for various purposes, soils, subsoil, should be approved by the Ministry of Health of the Kyrgyz Republic. This sanitary and hygienic principle of formation of standards has not yet been considered by the competent authorities of the Republic. Therefore, many projects are not able to ensure the protection of wildlife objects, and significant environmental violations are observed in them. MPC norms are poorly coordinated with each other. Approval of norms for new chemical substances is a rather long process. As a result, an increasing number of harmful substances enter the circulation without the basis of maximum standards for their content in different environments. Despite its shortcomings, the MAC is the only environmental criterion applied in design practice. It is clearly insufficient. Environmental rationing should be expanded and the issues should be referred to an independent section of the project.

In ensuring the environment favourable for people, animals and flora, control of its condition by means of environmental expertise, which is an assessment of the impact on the environment or environmental expertise of economic development projects, is of great importance. Of course, such expertise is not limited only to consideration of one or another production (economic) facility or approval of one or another norms. It is important to consider the objects related to the production unit, or the step-by-step implementation of the expansion of the composition of norms and standards, the development of methodological problems of standardisation, the improvement of the organisation of the technology of preparation of norms and standards.

At the same time, the enterprises, which are polluters of the environment, are not interested in the development of environmental protection activities: in the construction of new, improvement and maximum utilisation of any existing treatment facilities, in their rhythmic operation, as the costs (electricity, labour, etc.) for this type of activity fall on the cost of production. In this connection, there are facts of switching off the purification equipment (in particular, electrostatic precipitators) in order to save electricity, for which the personnel of the enterprises are financially encouraged. Cleaning is as if an unnecessary load on production, it is not planned and not stimulated, so the enterprises at the expense of it get a reserve for general saving and material incentives for other sections of production. Thus, when implementing the principle of cost minimisation in relation to environmental protection measures. For a particular economic object neither current nor prospective economic losses from irrational use of resources and environmental pollution are taken into account.

Waste is categorised into industrial, domestic and agricultural, toxic and non-toxic.

Disposal of production and consumption waste is carried out at:

- landfills for solid domestic waste, where some solid types of inert industrial waste may be disposed of in accordance with the established procedure.

Regular, systematic accounting of these aspects for all enterprises and farms becomes absolutely necessary in connection with the introduction of new methods for assessing the economic efficiency of investments and costs in environmental protection measures, as well as the transition to market relations.

In the process of management of the national economy, it is definitely necessary to provide a set of measures for the rational use of natural resources in the resource section, i.e. for each functional type of natural resources (water, air basins, forests, mineral resources). In our opinion, this subsystem should also include natural plant (fodder, food, medicinal), hunting, fish resources as such and their rational use.

The importance of the regional nature of environmental management should also be noted, since the optimal functioning of the economy in the future will be determined to a predominant extent by the level of rationality of resource use in the territorial context.

Management of natural resources is inconceivable without economic evaluation. The economic aspect of environmental protection measures consists in the rational choice of production process technology and technical means that ensure the implementation of necessary environmental protection measures at the lowest economic cost.

From an economic point of view, the preservation, restoration and transformation of the human environment in the direction necessary to ensure the life and needs of mankind requires the creation and improvement of appropriate productive forces to manage human interaction with the natural environment.

The economic aspect of nature management is based on economic potential, i.e. the aggregate ability of national economy branches to produce industrial and agricultural products, carry out capital construction, transport cargoes, and provide services to the population at a certain historical moment.[2]

Natural resources, contributing to the multiplication of national wealth, also require careful attitude to them. Of course, the use of resources has a pronounced territorial character. Therefore, we can also speak here about the use of resources in regional development. The utilisation of natural resources should be in general such that it contributes to the reduction of costs and increase of profit in social production. It is necessary to use the available raw materials and energy in the most efficient way.

The most common raw materials are mineral resources. Less common are organic raw materials - substances containing in their composition hydrocarbons or carbohydrates, their production compounds. All these types of raw materials are used after the appropriate

The special methods have been developed to improve materials related to social laws, economics, architecture, etc. Special methods have been developed to improve the materials related to social laws, economics, architecture, etc., as well as to improve the materials.

Kyrgyzstan has significant reserves of natural raw materials. For example, there are deposits of limestone, gypsum, coquina, quartz sands, brick and refractory clays, gravel, granite, etc. for the production of construction materials.

As it was mentioned above, waste utilisation is extremely important in the development of natural resources management measures. Practice has shown that on the territory of the Republic a large amount of various industrial wastes accumulates annually, including those with properties to fully or partially replace and ennoble natural raw materials.

The analysis shows that industrial wastes generated at the enterprises of the republic are not fully utilised. Fuel-containing wastes, which are often thrown into dumps, require the greatest attention. These include wastes from coal mining and preparation, ashes of TPPs. Thus, for example, coal mining wastes from "Almalyk", "Tash-Kumyr", "Kyzyl-Kiya" and "Kok-Yangak" deposits are expedient to use as a fuel-exchange additive in production of agloporite, expanded clay, bricks and drainage pipes. Charred rocks of heap heaps as light, coarse and fine aggregates in concrete production.

It is recommended to use ash from the Bishkek and Osh TPPs as an active mineral additive in the manufacture of cement and mixed binders, as well as an aggregate for heavy and lightweight concrete.

Rocks of mining enterprises are expediently used for construction of roads and hydraulic structures.

Smelting slags from Kadamjai and Hai-darkan metallurgical plants can be used for making aggregates of local binders, autoclave materials and slag-metals.

The metallurgical and machine-building enterprises of the Republic have generated about 100 thousand tonnes of foundry waste - "burnt earth", which is a good raw material for the production of finishing materials.

Cotton stalks, trimmings of fruit branches and vines, as well as waste from the textile and wood processing industries are suitable for use in the production of wall panels, dap, chipboard, sound and heat insulating materials.

Another aspect to be considered in the management decision-making process is the technology of manufacturing products. The main elements of technology are raw materials, apparatus, equipment, mechanisms and machinery. They are closely related to each other and are conditioned by the economy, the state and level of the environment and scientific and technological progress. Economic and environmental performance of production depends on the perfection of technology. The technology of production and energy production should be economically efficient and not exceed certain limits, after which it becomes impossible for resources to regenerate themselves. This requires rationalisation of production, its completeness, which implies minimisation of waste, elimination of losses, use of secondary raw materials.

In today's world, waste accounts for 96-98 per cent of the input material and only 2-4 per cent of the final product, according to various researchers. The challenge is to reduce waste by utilising it to the fullest extent possible. It is necessary to turn production into a closed system, developing low-waste and zero-waste forms of production.

Undoubtedly, increasing the efficiency of natural resources management is connected with economic measures - with wide introduction of resource-saving, low-waste technological processes, reduction of air and water pollution, in particular, creation of a closed-loop water supply structure for industrial enterprises, which is a system of water management of enterprises, under which all liquid wastes after appropriate treatment are returned for reuse and recycled into secondary raw materials. The use of production and consumption waste (secondary material resources) and the conversion of unused waste into used waste is of great importance for environmental protection, because it frees from neutralisation, landfilling or destruction of said waste, and also, as a rule, reduces energy and other costs, which in itself reduces environmental pollution and thus turns out to be economically beneficial.

Many phenomena in the process of natural resource management are associated with inappropriate interference in the ecological system, as it can sometimes create a false impression of well-being. For example, the widespread use in agriculture of mineral fertilisers, pesticides, pesticides, pesticides, biostimulants can at first give an explosive effect of yields and productivity, but over time the ecological system is depleted, which has a negative effect on increasing their ability to form stable annual crop yields.

A similar situation can be found in the ploughing of agricultural land, where wind and water erosion is formed, which contributes to the appearance of landslides, gullies, and, consequently, causes a decrease in crop yields and animal productivity.

That is why man in his activities cannot fail to observe the principles and rules of ecology, for, having once stepped over a single phase of the sequential development of an ecological system, or having ignored at least one of its living components, clearly adapted to a certain ecological niche, one can break the whole system, which is very difficult, and sometimes even impossible, to restore. This applies to all components of an ecological system - be it animal, plant, microzone, terrain exposures, moisture, windiness, solar radiation, etc.

Naturally, the management of natural resources cannot act in an isolated world, but is an integral part of the management of the entire national economy. In this regard, the comprehensiveness and optimality of management decisions is of certain interest. The principle of comprehensiveness and optimality, unfortunately, is not yet sufficiently used in Kyrgyzstan. For example, in the coal industry, waste dumps could be used as construction materials. However, in the existing economic practice, many extractive, processing, manufacturing industries focus mainly on the raw materials and components in them and the products produced from them. Therefore, enterprises that were insufficiently prepared and not oriented to full, deep complex extraction and processing of raw materials were often put into operation.

In market conditions, the role of scientific and technical progress in nature management should be noted. However, in Kyrgyzstan the available scientific and technical achievements in the field of rationalisation of natural resources use, in many cases locally implemented in practice, are not used on a large scale, which significantly reduces the efficiency of resource availability of the entire national economic complex.

3. Criteria and principles of natural resource management

The assessment of trends in the organisation of natural resource use both in our republic and in other former Soviet republics makes it possible to identify and trace priorities in management decision-making. First of all, it is necessary to note the close interrelation of the problems of nature management and its programmes and concepts with social issues. This is understandable, since everything that is done is directly derived from the essence of human activity, which is directed for the sake and in the name of man. As T. Ivanova and N. Golovatskaya correctly pointed out, we are talking, first of all, about the protection of public health and the orientation of nature management towards people. The state of health depends on many factors. Firstly, health is greatly influenced by the volume, structure, quality and standards of consumption, housing, organisation of everyday life and leisure, cultural development, and the degree of socio-political comfort and democracy of society. Secondly, the state of the environment and labour conditions, i.e. the level of development of the entire material and technical base of production, are reflected in the health of society. Thirdly, positive and negative processes in society and the natural environment determine the health of future generations. Fourthly, the state of public health depends on the efficiency of functioning of social institutions and, first of all, health care.[1]

Acting as the main criterion, human health should be enshrined in legislative acts related to the use of natural resources and environmental protection. For example, the Water Law of the Kyrgyz Republic requires that the rational use of water should be ensured during the location, design, construction and commissioning of new facilities, as well as during the introduction of new technological processes affecting the state of water, provided that the drinking and domestic needs of the population are met as a matter of priority. According to the Law, measures should be implemented everywhere to ensure that water withdrawn from water bodies and returned from them is accounted for, that water is protected from salinisation and pollution, littering, that harmful effects of water are prevented, that flooding of land is limited to the minimum necessary extent, and that favourable natural conditions and landscapes are preserved. In addition, measures ensuring protection of fish, other aquatic animals and plants and conditions for their reproduction should be implemented in a timely manner at water bodies.

At the same time, it should be borne in mind that numerous opportunities to improve the efficiency of water use and protection should not be associated with high costs.

Another group of requirements related to the problem of human health and rational environmental management is the widespread elimination of existing and emerging sources of negative impact on the environment, i.e. we are talking about rational and rules, avoidance of pollution, fulfilment of special requirements. As an example, the following can be given. According to the Legislation, all enterprises are obliged to use the natural resources provided to them, including land, water, minerals, forests and others, rationally, i.e. carefully, comprehensively and in compliance with environmental protection standards, in the most efficient and strictly, for their intended purpose. In particular, they shall not allow pollution and contamination of land plots under their use when carrying out works related to disturbance of the land surface, preserve the fertile soil layer and restore the condition of the territory in a timely manner. The Law "On Subsoil" of the Kyrgyz Republic obliges subsoil users not only to ensure their rational use and preservation, but also to protect air, land, forests, waters, nature reserves, natural monuments and other objects of the natural environment, as well as buildings and structures from the harmful effects of work related to the use of subsoil.

Enterprises are obliged to build, equip and properly operate wastewater treatment facilities. They may emit emissions into the atmosphere and water bodies only with permits from the authorities authorised to regulate the use and protection of the atmosphere and water bodies. Any emissions must not result in exceeding the established norms of maximum permissible concentrations (MPC) of pollutants in the atmosphere and water bodies.

In the same row is the fight against noise. It is a systematic development and implementation of a set of engineering, technical and organisational measures aimed at reducing noise levels to sanitary norms. Fulfilment of legal requirements, norms and specific tasks on environmental protection are constantly monitored by authorised state bodies, and their violation is prosecuted by law. For example, sanitary and epidemiological authorities have the right to prohibit or temporarily suspend the operation of existing facilities that may cause harm to human health through emissions, waste or refuse. In addition, the system of environmental management in the Kyrgyz Republic is as follows: subsoil is controlled by the State Agency for Geology of Mineral Resources, land by the State Agency for Land Construction and Land Resources under the Government of the Kyrgyz Republic, and water by the Sanitary and Epidemiological Service. There are various kinds of fishery and hunting inspections. There are meteorological, seismic stations and others. The Ministry of Emergency Situations and Civil Defence is responsible for measures against geological hazards. A reasonable concentration of forces is necessary here.

As a generalising indicator for measuring nature management in terms of its interaction with humans, we can propose the amount of damage to health and the environment from the violation of the economic balance of nature. Of course, it is a difficult task to quantify the amount of damage to health and the environment. Nevertheless, to a first approximation, such damage depends on the volume and level of utilisation of resources involved in economic activity. Direct losses from unrealised waste reduce the efficiency of production. At the same time, they cause an increase in morbidity of the population and disturbance of ecological balance in nature, destruction of cultural monuments, landscape, etc. At the same time, in addition to traditionally accounted losses and waste of raw materials, materials, energy, there is also noise, vibration, dust, poisonous gases, residual pressure, heat and other types of energy extraction.

Economic damage caused as a result of natural phenomena and processes is difficult to analyse and economically assess. In a number of cases, so-called compensatory measures of social protection are used. For example, part of the damage was compensated to residents of Tosoi and Komsomol villages who suffered as a result of a landslide in 1994.

Air pollution has a certain impact on human health. Air pollution problems are localised and are usually caused by low quality fuels used in power plants and transport.

Sanitary and hygienic conditions, consumption of contaminated food, poor water supply, poor quality of public infrastructure, etc. are to some extent responsible for the morbidity.

Therefore, the preservation of human health must play a central role in conservation and natural resource management.

The next management criterion is socio-economic growth and development. There are undoubtedly many important challenges facing Kyrgyzstan, and it is necessary to conduct a thorough analysis to identify those natural resource management issues that need to be prioritised, given that among the main objectives is to improve the living standards of the population and reduce poverty.

In this regard, land use in the context of crop production is a typical example. Crop production is an economic problem. The existing deficiencies in land utilisation, slow structural reforms in agriculture, delayed payments to grain producers and lack of credit facilities are the main reasons for lagging behind in the agricultural sector. As a result, many farmers have shifted from producing crops for sale to producing them as a means of subsistence. This has led to the fact that the volume of sold products has declined more significantly than the main volume of agricultural production.

The above-mentioned problems of decreasing agricultural productivity are mainly related to improper land use, insufficient water supply to irrigated lands and lack of sufficient fertiliser needed to achieve a given yield.

In other words, sound agricultural economic policy will not only have a major impact in addressing rural economic problems, but it is also of paramount importance in addressing specific conservation problems. This is because rural economic problems often, while causing farmers and livestock keepers to respond appropriately to poverty and stability, also address conservation problems at the same time.

When considering the criteria and principles of natural resources management, it is necessary to note the exceptional importance of environmental protection activities and effective use of environmental infrastructure. Thus, as studies show, at present, economists have no doubts that the costs of environmental protection from pollution, reproduction of natural resources and restoration of environmental quality (in case of its reduction in certain areas) are socially necessary. However, the criteria for determining such costs in the medium and long term have not been developed, and there is no mechanism for realising these costs in economic activity.

So far, in modern economic practice, there is often resistance to realising the environmental costs of planned measures.

Even specifically allocated capital investments for these purposes are in many cases significantly underutilised. Preventive measures and corresponding technological redesigns at production facilities, including those aimed at deepening the integrated processing of natural resources, increasing the full use of raw materials and basic materials, and mobilising and making fuller use of secondary material resources, are often not introduced.

Or an example of how many environmental infrastructure facilities are sometimes used inefficiently. For example, canals are cleaned much less frequently, water control equipment is not repaired, and as a result canal capacity and water distribution volume have decreased.

All this calls for greater attention to the efficient use and maintenance of existing fixed assets, especially in the area of environmental infrastructure.

Of course, as a criterion in natural resource management, the assessment of land and natural resources in terms of the impact of their management on the introduction of market economy, on human health and the environment.

Assessment of land from the point of view of environmental protection and human health should be carried out taking into account the ecological and economic principle of economic functioning, which is applicable especially when natural resources have become less and production waste has become more, and which is based not only on economic and engineering calculations, but also on ecological and technological ones. This is especially important in the production cycle of the agro-industrial complex, where living objects: plants and animals are used as means and objects of human labour. The basis in this case are norms and maximum permissible parameters of anthropogenic and technogenic impact of a particular type of production and products on the ecological balance of nature. In the process of any production it is necessary: not to exploit natural resources until their complete depletion; not to interfere with the natural course of the biosphere's cycle of substances; to choose a technological method of production that would not harm the life forms of the surrounding nature.

As in other spheres of activity, there is a strong need to introduce a market mechanism in the use of natural resources. However, this process is held back for a number of reasons. This can be traced on the example of agricultural land use, which on the one hand "supposedly" has a price and on the other hand there is an equalisation in the use of land of different quality.

The next criterion for assessing manageability is the assessment of the efficiency of replacing primary resources with waste. The main principle here is the full use of material mass withdrawn from the natural environment, which should give the greatest effect in the process of economy of living and embodied labour. However, despite the obvious effect that can be obtained from increasing the level of rationality in the use of material primary and secondary resources, some material resources are not used or are used irrationally. One of the main reasons for this is insufficient development of the methodological apparatus for determining the effectiveness of interindustry self-financing mechanism and sufficiently effective methods and forms of stimulating the involvement in circulation of industrial waste, secondary material resources, increasing the complexity of the use of raw materials; imperfect pricing in this area.

Assessment of qualitative characteristics, physical and chemical properties, consumer value of many mass production and non-production wastes generated in the national economy shows that wastes have a very high consumer value, allowing them not only to be identified with the corresponding primary resources, but also to replace them (compensate) with significant economic advantages. This Constantia is based on many years of both domestic and foreign economic practice in the mobilisation of such wastes as scrap and waste fibrous materials, ferrous metallurgy slag, ash from thermal power plants, food waste, car tyres, glass sludge, plastics, used machine oils, etc.

Rationality of resource use in this case is determined, on the one hand, by the creation of various consumer values from waste, the quality of each of which would meet the requirements of society, and on the other hand, by the amount of damage to human health and the environment that these wastes can cause.

To implement the principles of natural resources management, it is necessary, first of all, to have a good information base, which allows analysing the state of the environment and human health, and providing all management bodies and the population with the necessary information.

Of course, the implementation of all the principles of manageability is possible if there are appropriate means and resources, as well as conditions (prerequisites) for free and prompt redistribution of resources between different spheres of activity, including non-productive spheres.

The second group of prerequisites on which the management principle is based is the elimination of departmental barriers to the implementation of effective environmental protection measures by regions.

As in any management system, it is also necessary to have a clear distribution of functional responsibilities between different management bodies and executors. After all, much depends on the extent to which powers and responsibilities are balanced and defined when making managerial decisions. Economic methods of management should be favoured, because administrative methods of management, in fact, being command and order, limit the freedom of action.

It is known that the efficiency of a management system is higher the more promptly management decisions are made, taking into account the time factor.

In the implementation of any activities, the means and resources attracted for these purposes play a decisive role. However, in the Kyrgyz Republic, such an opportunity is almost non-existent, as funds are not available in sufficient quantity and we have not yet developed an economic mechanism that would facilitate the wide use of both the existing potential and the attraction of investments for these purposes.

It is also necessary to take into account a certain decentralisation of management, which will have a positive impact on the protection of health and environment with the unconditional presence of republican management bodies, especially in the implementation of structural reforms and the development of fundamental research. Of course, at the same time, the management should be democratic and flexible in the transition to market relations.

Organisation of measures on environmental protection and rational use of natural resources should be based on knowledge and deep scientific application of the laws of development and functioning Laws of Nature. On the basis of these factors, society will be able to systematically manage natural processes on regional and global scales and create special mechanisms to guide the development of the biosphere in the direction necessary for man.

As noted above, it is about the content of environmental protection to ensure control over the state of the environment and sources of its pollution on the basis of the use of the latest scientific and technological achievements and means, to introduce new effective methods and systems for the development of mineral deposits, to combat harmful emissions of substances into the atmosphere, industrial, transport and other noise, vibrations, electric and magnetic fields and radiation, to implement measures for the integrated and rational use of natural resources, to ensure the protection of the environment, to ensure the protection of the environment and the sources of its pollution. We propose the following structure of economic problems of natural resources management.

CHAPTER 2

Current state of national natural resources management

1. Assessment of natural conditions and resources involved in economic activities

In order to justify management decisions in the field of nature management, especially in the development of comprehensive economic development programmes, it is necessary to organise a system of qualitative assessment of natural resources and natural conditions involved in economic activity.

As studies show, natural resources have a great impact on the expenditure of social labour required to create the economic potential of the state. In addition to natural resources, it is necessary to assess the natural conditions, as they create the possibility of human activity and production. These include solar radiation, internal heat of the earth, geographical position of the area, relief, climate, precipitation. At a certain stage of development of productive forces, natural conditions become resources, for example: solar radiation, wind, sea tides, internal heat of the earth.

Natural resources are elements and forces of nature that can be used in the production and non-production spheres to meet the needs of people. Natural resources include elements of the lithosphere, hydrosphere, atmosphere: minerals, soils, rivers, lakes, underground water, seas and oceans, flora and fauna.

Production, developing, uses all kinds of natural resources and their differentiation, and the development of production is ultimately a process of interaction between society and nature. Thus, the geographical environment affects man through the production relations arising in a given area on the basis of these productive forces, the first condition for the development of which are the properties of this environment.

To the properties of the environment, which have a great influence on the development of production, K.Marx included the difference of natural conditions, promoting the division of labour and economic ties between people.

For both natural conditions and natural resources, proper accounting is necessary.

Unfortunately, there is no proper accounting of natural resources in the Kyrgyz Republic. Therefore, many issues of natural resource use today remain unclear. In our opinion, it would be advisable to introduce in practice a special form of accounting of both production consumption of resources and waste generation and waste compensating resources (Tables 1-3).

In the presence of such systematic accounting there is a real possibility of forming regional and national resource interbranch balances of primary (natural) resources and wastes of production and final consumption, and on their basis there is an opportunity to develop schemes of rational resource use and target programmes of resource use. In such programmes, the need of the national economy in this or that material substance should be provided according to the optimal structure of sources of its obtaining with extremely full and comprehensive use of the mass of exploited at all limits of the technological cycle, stages of production and consumption.

The new ecological situation, in which elements of market economy have to function, makes it necessary to consider economic problems from the angle of requirements of nature management and nature management from the angle of requirements of economic development. These systems can be local, regional and global, depending on which part of their interactions becomes the object of scientific analysis.

Local economic-ecological systems are those where a separate area of economic activity (enterprise) is considered in relationship with the nearby environment or its separate objects - air or water basin, subsoil, landscape.

Regional economic-environmental systems are characterised by a higher level and diversity of economic objects and a correspondingly wider spatial extent of the environment itself. They include systems where the components are a large group of enterprises and various environmental objects. The Osh and Jalal-Abad oblasts of the Kyrgyz Republic, as historically formed, can be referred to the region. Taking into account the geopolitical peculiarity, it is called the South of the Kyrgyz Republic.

The global economic-ecological system is a system where, on the one hand, the economy as a whole is considered and, on the other hand, the entire environment is considered, either on a planetary scale or within national borders. The territory of the Kyrgyz Republic can be referred to the latter category of the economic-ecological system.

The intensive interpenetration and merging of the economy and ecology into a single system at the local, regional and global levels, when natural resources simultaneously act as an important element of both ecology and economy, generates complex changes in the state of nature and economy, and largely determines the nature of their interactions. The emerging problem requires new research and management methods based on the knowledge of the laws of existence, sustainability and development of complex ecological and economic systems.

Management of the economic-ecological system, as well as any management, includes as mandatory the following three elements: the object of management, means of management and the purpose of management.

The object of management is understood as the planetary cycle of substances and energy in the geographical shell of the Earth. Part of the links of this cycle is subject to the laws of development of productive forces. Another part is subject to the laws of development of dynamic systems seeking to move from a state of disequilibrium to a state of equilibrium.

Means of economic and ecological system management in market conditions are economic levers: economic evaluation of resources from the national economic standpoint; payment for pollution and, in general, deterioration of the quality of the environment and individual resources, for exceeding the norms of their consumption; income from the use of natural resources and the environment as a whole.

The goal of economic-ecological system management is to increase the economic efficiency of production. This criterion is understood as such an exchange of substances and energy (in the process of their cycling) between production and the environment, which is carried out on a scale to meet the needs of society.

The solution of problem management of environmental management, i.e. environmental protection and rational use of natural resources is a process of drawing up a scientifically based concept and for practical use of the programme for selection of criteria, their evaluation and obtaining optimal solutions.

At the same time, natural resources, the natural environment and the conditions of their functioning are not only an object of economic activity, but also an ongoing factor in the all-round development of man.

Natural resources and natural conditions of Southern Kyrgyzstan are diverse. Let us consider three directions of research. First, the results of research allow us to determine the permissible loads on humans, the natural environment and the economy. It is necessary to determine a set of equilibrium states of the ecological and economic system of the region. The issues of stabilisation and balanced development of all components of the system are also important here. The equilibria of ecological and economic systems considered separately are extremely complex and have not been studied to the end. The limits of adaptability and self-healing in many cases are unknown, and the problem of sustainability of ecological and economic systems is to be posed, theoretically substantiated and ways and means of realising practical conclusions are to be outlined.

The second is the development of a scientific basis for forecasting, regulation and management.

An important place is occupied by the study of formation and development of territorial natural-economic complexes in river basins, zoning of territories taking into account both economy and natural environment, their balanced development.

The third is the assessment of natural resources and the properties and quality of the environment. The south of Kyrgyzstan is the largest administrative region of the Kyrgyz Republic. Its area is 73.9 thousand kilometres or 37.2% of the country's territory.

In the conditions of transition to a market economy the problem of utilisation of this territory, protection and reproduction of natural resources arises.

The scientific methodology underlying such a problem should be based on the prospective target designs of territorial development, on the basis of optimal use of available regional resources and taking into account both comprehensive inter-sectoral relations and multiple complex bilateral relations of production with the environment.

Most of the territory of the region consists of ridges and foothills, and the smaller part - of intermountain valleys. The region is located at the junction of the greatest mountain systems - the Tien Shan and Pamir-Alai, has sharply pronounced vertical zonality within the range from 5000 to 7000 metres above sea level and has a multi-stage character. The river valleys in the margins of the Fergana Basin are characterised by relatively flat relief and complex forest cover. In some places there are isolated elevations, for example, Mount Suleiman-Too (Osh). The latter has a twofold significance. On the one hand, the rock strata are valuable as suitable for the production of limestone, lime and torn stones. The relief allows to organise here a recreation area and tourist routes. On the other hand, Mount Suleiman-Too should have ecological inviolability as a natural monument.

In the south of the region there is the high-mountainous Alai valley, in the north - the large intermountain depression Ketmen-Tyube and in the north-west - Chatkal valley. They require solving the problem of optimising the interaction between the economy and the environment. In our opinion, communication and recreational sides of this serious problem are of special interest.

The Adyrs directly surround the Fergana Valley. They are winding-wavy plains, strongly dissected by numerous ravines. Their absolute elevations usually do not exceed 1000-1500 metres above sea level. In economic terms, adyrs with lower location are insufficiently used for rainfed crops, as well as for cattle grazing in spring and autumn.

Behind the Adyrs there is a strip of high foothills and waders of the main mountain ranges. The most elevated part is the Zaalai Range, which belongs to the Pamirs. This huge, eternally snow-covered ridge, which reaches a height of 7134 metres at the highest peak in Lenin Peak, closes the Alai Valley from the south, which is of great economic importance. It is used as pasture by the Alai and Kara-Suu districts and leased by districts in the neighbouring republic of Uzbekistan. Its northern slopes are located in Kyrgyzstan and its southern slopes are located in the Gorno-Badakhshan Autonomous Oblast of Tajikistan.

The description of the geographical location of the region, its climatic conditions is necessary not only for drawing up complex balances of schemes for the rational use of natural (material) resources, taking into account the possibilities of marginal mobilisation of public production waste, but also for drawing up territorial complex schemes for nature protection. Such schemes in the future should be created and dictate the redistribution of investments between production on primary resources and wastes in extractive, processing, manufacturing industries of material production, in particular, in the production of metals and alloys, chemicals and materials, construction materials, in agriculture, food, textile industry, etc. The determining factor of the priority of involvement of a particular resource in the turnover should be the indicator of the level of total labour intensity of its use or processing. Based on these values, the structure of the base, the nomenclature of exploited sources and types of raw materials in the coming period should be determined.

For example, natural conditions will influence the construction process and construction products to a different extent. Regarding the South of the Republic we can say that most zones of the region are characterised by the most favourable conditions for construction, although there are such areas as Alai-Kuu and Chon-Alai, Chatkal, etc., where climatic conditions are almost equal to those of the Far North.

In connection with the industrial development of hard-to-reach and remote areas, the problem of supplying industrial machinery and technological equipment, which has not been solved so far, becomes especially urgent in the perspective period. The problem of finding free territories for construction is becoming more and more acute for the development of the cities of the republic, taking into account that flat territories account for only 7 per cent of the total area. The issue of rational use of worked-out mine workings (the fund of which amounts to several million cubic metres), such as ravines, quarries and mines, has become acute. This will make it possible to effectively solve the most important national economic tasks of increasing the efficiency of land fund use, heat energy economy, environmental protection and health improvement, improvement of architectural and aesthetic appearance of building of cities and settlements, increasing safety, convenience and speed of transport and pedestrian traffic at the same time, civil defence tasks. In order to reduce the use of valuable agricultural land for construction, the problem of using complex terrain (sloping areas) for the development of settlements is currently and in the future. Natural and climatic factors have an important impact on the solution of urban planning problems: types of settlements and dwellings, their planning and development. The south of Kyrgyzstan belongs to the III and IV climatic sub-areas. These areas are characterised by dry and hot climate, high dustiness of the air (during 30 days a year dust concentrations of 1.5 m /m[23]).

Here, a special requirement is made for the correct orientation of houses in the latitudinal direction. In the Fergana Valley, climatic conditions allow construction throughout the year, which affects the lower estimated cost of facilities compared to other zones of the country. For high-mountainous areas it is characteristic to reduce labour productivity up to 20% from the plain conditions of construction, and in the winter period in the mountains it decreases up to 10% in comparison with the summer period.

High temperatures, strong cyclical temperature differences, low relative humidity and intense solar radiation affect the technology of construction production.

A significant share of construction costs in seismic areas is the cost of antiseismic measures. They are approximately 4,8 and 12% of the estimated cost for 7,8,9-point design seismicity instead of 5-8%, provided by the standards of specific capital investments. The city of Osh is located in the centres of possible earthquakes.[1] The construction of the Papan reservoir on the middle reaches of the Ak-Buura River has led to changes in the natural and climatic conditions of Osh and the neighbouring Kara-Suu and Aravan districts. The possibility of landslides, snow avalanches, mudflows requires specific organisation and technology of works, additional costs for territory development.

The problem of population location is an urgent environmental challenge. On the one hand, the development of new areas requires very serious expenditures on transport, roads, commercial, cultural and medical services. Reconstruction of the housing stock must be carried out in accordance with master plans that provide for the demolition of non-seismic and other geohazardous objects. In addition, plots of fertile land are occupied for development

On the other hand, the South of Kyrgyzstan is very unevenly populated. The average population density of the region is 20 people per square kilometre. However, this indicator does not reveal the current picture of the distribution of inhabitants. This is a consequence of the complexity of the relief, orthographic structure and varying degrees of economic development of the population.

The most densely populated areas are the foothills, pre-Fergana plains in the Osh-Karasu zone, Kugart, Kurshab, Kara-Urgur and Shaidan valleys, i.e. zones of intensive irrigated agriculture. In some parts of this zone, for example, in Osh-Karasui oasis and Kugart valley, the population density per 1 square kilometre reaches 70-100 people.

The foothill adyr strip and sloping slopes of the lower parts of the ridges are characterised by low population density. Small villages are located mainly in intermountain depressions with agricultural massifs. The high-mountain intermountain valleys of the Alai, Chatkal, and Alai-Kuus valleys are sparsely populated, with a population density of 1-2 people per 1 square kilometre.

If we visually present the modern network of settlements in the region, we get the following picture: around the Fergana valley in the zone of intensive irrigated agriculture, as well as in the lower parts of large valleys are located the cities of Osh, Jalal-Abad, Uzgen, Kara-Suu, Kyzyl-Kiya and most of the rural settlements, large both in terms of population and the functions they fulfil. The largest of them are the villages of Aravan, Eski-Nookat, Nariman, Kashgar-, Kashgar-, and Kashgar-.
Kyshtak, Suzak, Bazar-Korgon, Kochkor-Ata, Sovetskoye, Ala-Buka, Gulcha, Mady, Karavan, each with 1.5 to 20 thousand or more inhabitants.

Above - in the adyr zone and foothills - most urban settlements of Aidarken, Kadamjai, Sumsar, Shakaftar, Myrza-Ake, Kara-Kulja, Isfana, as well as a large number of other settlements are located. Many settlements, relatively small in size, are located along narrow river valleys, in the zone of nut-bearing forests. High up in the mountains, settlements are small and numbered in units.

The economic well-being of a country is unthinkable without a properly organised system of recreation of people. This raises the problem of studying the natural conditions of the territories adjacent to industrial facilities in order to select the best places for recreation.

Creating the best possible conditions for recreation, which provides for a longer stay of a person in the fresh air not only in summer but also in winter, is a state task.

State measures to protect the health of the people are based on the principle of prevention and prevention of disease development. An important role in this is played by the geographical environment, natural and climatic factors of resorts and sanatoria.

Their successes spa prophylaxis and therapy owe to the achievements of climatological science in the study of physical and chemical properties of the environment and the influence of its individual components on the healthy and sick organism.

The climate of Kyrgyzstan is the most active biological highly effective therapeutic factor.

In Southern Kyrgyzstan, on the slopes of Chatkal and Fergana ranges, at altitudes from 1400 to 2400 m, there are Arslanbob and Sary-Chelek unique nut and fruit forests (their area is more than 300 thousand ha).

Mountains, forests, mountain-marine (with the emergence of reservoirs), mountain-forest climate, mountain rivers and mountain lakes are the greatest gift of nature, they bring people great material benefits and health. Mountain tourism and mountaineering ascents, medical treatment and recreation are invaluable sources of health. However, they are currently not used rationally. In the South of Kyrgyzstan there are Jalal-Abad resort, Osh recreation centre, sanatorium "Automobilist", "Cooperator" and others.

Jalal-Abad resort is one of the oldest and most popular resorts in Central Asia. It is located in the north-eastern part of the Fergana Basin on the western slopes of the Ayuu-Too Mountain (Fergana Range) at an altitude of 975 metres above sea level. Due to its high location, the resort overlooks the vast, picturesque Kugar valley, located 300 metres below the resort level, closed from the north by the snowy peaks of Arslan-Bob. The lower tiers of the mountains surrounding the valley are covered with pistachio and walnut forests; the Kugarst river flows in the valley, originating from glaciers and is a tributary of the Kara-Darya.

The therapeutic properties of the mineral springs of the resort have been known since ancient times. Currently, a bottling of Jalal-Abad mineral water has been organised. Consumption of this valuable water is recommended to improve digestion and prevent gastric and kidney diseases due to the chemical substances it contains that the human body needs (Table 5).

Chemical composition of Jalal-Abad mineral water

Table 5

Cations	mg/l	Anions	mg/l
Sodium	633	Hydrogen	0
Calcium	79,5	Phosphates	0,22
Magnesium	70,4	Sulphates	1310
Strontium	5,51	Nitrates	1,13
Lithium	0,033	Chlorides	518
	30	Fluorides	0,72
Iron (total amount)	0,269		
Manganese	0,5		
Ammonia			

The content of chemical elements in water, as doctors' recommendations show, has a favourable effect on people's health and it also implies a great economic benefit when skilfully organised and implemented on a mass scale.

Similarly, but on a larger scale, can be said about other sources and places of recreation. Thus, the Osh rest house is located on the elevated eastern side of Osh city, at an altitude of 1175 m above sea level. Osh rest house belongs to the low-altitude with semi-desert climate (hot summer, mild winter), the average monthly temperature of summer period fluctuates within 220C (June), 250C (July). The maximum temperature reaches 390C and more in July. Constantly blowing cool winds along the Ak-Buura River cool the air. The minimum temperature in January is up to 200C. The clearest and sunniest days are observed in summer and autumn, when precipitation is rare. In the floodplain of the Ak-Buura river there are sanatoria "Automobilist" and "Cooperator". Their climatic characteristics are the same as those of the Osh recreation centre.

These entities can serve to organise a strong tourism industry in Southern Kyrgyzstan and serve to increase the economic potential of the region and the republic.

One of the resources involved in economic activity is mineral raw materials for mining and metallurgical industry. At present, the industry provides about 15% of the total industrial output of the South of the Republic. The products produced here are antimony and its compounds, mercury.

The production of antimony and its compounds, as well as mercury, is accompanied by air pollution, which has a direct impact on human health. Antimony production facilities in Kadamjai and mercury production facilities in Khaidarkan are located in the centres of villages, each with approximately 12,000 inhabitants. The environmental problem here is that there are no sanitary or safety zones separating housing from the mines.

The situation is exacerbated by the fact that mining operations often neglect requirements related to environmental protection, safety and occupational health.

The Khaidarkan and Kadamjai Combines alone have generated millions of tonnes of waste, which contains hundreds of thousands of tonnes of mercury and antimony compounds, arsenic, fluorites, heavy and base metal compounds, and radioactive contamination from uranium and radium.

Air pollution is not only related to the content and properties of natural resources involved in economic activity, but also to a greater extent to the nature of processing and the state of industry itself. In the South of the Republic, most of the industrial enterprises use outdated technology and equipment, the pollution control devices installed at the enterprises are old and do not meet modern requirements. Due to heavy loads and poor condition of the instruments it is not possible to fully record emissions and pollution levels. Some wastes, such as those containing mercury, are disposed of in municipal landfills, while others are left unattended in various places.

Energy resources account for a significant part of the resources involved in the economy. The main energy resource in the Republic is coal, with reserves of about 2 billion tonnes. Historically, most of the coal is extracted in the South of the Republic. Coal is a good source of heat and its average thermal value is 4000 kcal/kg and its sulphur content is less than one. Ash content varies from 10 to 35%, depending on the content of extracts[1] .

For at least thirty years, the planned economy focused on increasing production and quantitative indicators, with little regard for costs, and the prices of natural and capital resources were grossly underpriced. The development of heavy industry, where coal was often the main source of energy, remained a constant priority. From an environmental point of view, industrial development on this scale, wherever it occurs, is destructive to the environment. But in countries with centralised economies, air pollution was further exacerbated by the fact that energy and raw materials were used in a highly unsustainable manner because they were undervalued.

Consumption of coal on a large scale entails an increase in the emission of dust and gases into the atmosphere. Thus, in 1990 the Kyrgyz Republic produced 4202.0 thousand tonnes of coal; 24.0 thousand m^3 of firewood; 218.7 thousand tonnes of oil and 131.0 million m^3 of gas. These indicators in 1995, although sharply decreased, amounting to 12.4 thousand tonnes of coal, 21 thousand m^3 of wood, 88.5 thousand tonnes of oil and 35.7 million m^3 of gas. However, they also have a certain impact on human health.[1] The biggest impact on the environment and health will be dust emissions, especially at TPPs and boiler-houses, coal-fired power plants with poorly working dust collectors. Nevertheless, the increase of coal production to cover the needs can be achieved by involving small deposits in open-pit mining and creating new coal mining complexes on the basis of newly explored deposits. On the territory of the South there are more than 30 studied deposits and coal occurrences, which can be developed by the open-cut method, which does not require large capital investments. The largest coal deposits are located in the Uzgen coal basin, in Tash-Kumyr, Kok-Yangak, Kyzyl-Kiya, Suluk-Ta, Almalyk.

In the South of the Republic, oil and gas production is also involved in economic activity, though still in small amounts (about 20 per cent of the Republic's needs). The presence of oil and combustible gas deposits was established as early as 1901. In Jalal-Abad oblast oil production is about 100 thousand tonnes per year and natural gas production is 30 million cubic metres per year. Here, too, environmental protection measures are not carried out at the proper level. All works are carried out without preliminary environmental expertise. The new enterprise does not have an environmental passport.

It is known that the Kyrgyz Republic has huge hydroelectric resources. There are 5 large and 13 small operating hydroelectric power plants with a capacity of 3000 mv. The problems here are related to the fact that in the past the costs related to environmental protection during construction of HPPs were not fully taken into account.

Most of the large hydroelectric power plants (HPPs) are built in the South of Kyrgyzstan. These are Toktogul, Kurpsai, Tashkumyr, Uchkorgon, Shamaldysai HPPs. Construction of the Kambarata HPP has begun. In the past, the energy sector developed according to the plan determined from the Centre and was practically unrelated to the local economy. This includes flooding of most of the fertile land, disturbance of the biological balance, disappearance of animal and plant habitats in the places of HPP construction, etc. This shows that along with the construction of HPPs, there is a need to improve the quality of life in the region. This shows that along with the undoubted benefits, it is necessary to take into account the negative consequences of HPP construction, primarily for the environment and human health.

As noted above, the territory of southern Kyrgyzstan is rich in deposits of natural stone (marble, granite, decorative limestone and other rocks), which can be used to obtain facing materials and which are in wide demand everywhere and can be a source not only for domestic consumption, but also for export to near and far abroad.

So far there is only one explored deposit of limestone-shellstone in the region (Sarytash deposit), developed by the Oshstromaterials Association (Aktash JSC).

In 1992, exploration of the Akart white marble deposit was completed. Part of the reserves of this deposit are used by the Southern Kyrgyz Geological Expedition for the production of facing tiles with a capacity of 10 thousand square metres per year.

Analyses show that the reserves of the deposit allow to increase the extraction of stone many times for processing into finished products or selling as raw material. For this purpose appropriate investments are required.

In 1992, the South Kyrgyz Geological Expedition planned to complete exploration of the Adzhike anhydride deposit and conduct geological exploration at the Chon-Buloli grey granite deposit. Both objects are located in the Alai district and are characterised by favourable mining and economic conditions. However, due to lack of funding, work on both sites has been mothballed.

In addition to the above-mentioned deposits, there are a number of other promising manifestations of marble in the basin of the Lyalyak River (Lyalyak district), on the left bank of the Sokh River (Kadamjai district), granites (Gaz intrusive massif), pictrites - mountainous eruptive rocks (Karachatyr zone in Naukat district, Sartalinsk strip to the north of the village of Khaydarkan, etc.), pink marble Eruchey Taldysu in the area of the Taldyk pass, etc.). In the area of Taldysu Taldyk Pass, pink marbles of Eruchey Taldysu, etc.).

The natural conditions and resources of Southern Kyrgyzstan create favourable prerequisites for its economic development. Among mineral resources, first of all, there are fossils of fuel and energy importance - coal, oil, gas, and deposits of non-ferrous metals. Coal, oil and gas industry and non-ferrous metallurgy have been created on their basis. However, from year to year there is a shortage of combustible gas in industry and in everyday life.

The Isbaskent-Maili-Sai-Osh-Sokh-Khaidarkan gas pipelines are currently in operation. Osh oblast is a leading CIS region in antimony and mercury production.

At present, 592 deposits of construction materials in the south of the republic have been evaluated to various degrees, of which only 30 are exploited. Information on the state of development of commonly occurring deposits of construction materials, facing stones and other non-metallic minerals is given in Table 6.

The use of so-called non-traditional types of mineral raw materials in the agro-industrial complex may be of practical interest from the point of view of state regulation: neoliths, bentonite, glauconite and polygorskite clays, talc-magnesite rocks of trepels. There are practically unexplored manifestations of phosphorites and phosphate-carbonate raw materials.

In recent years, the country has experienced a decline in the yield of fruit, horticultural, technical and other agricultural crops due to an acute shortage of combustible sulphur to combat agricultural pests.

According to the Ken-Too Design and Research Centre, the republic's total demand for sulphur, including the need for sulphuric acid, is about 7,000 tonnes annually.

In the south of Kyrgyzstan, the Chang-Gyr-Tash nugget sulphur deposit in

Suzak district is known. In 1993, the Southern Kyrgyz geological expedition in this deposit selected the most favourable local area for open-pit mining. However, sulphur reserves at this site are only 5000 tonnes. Thus, the problem of sulphur supply in the republic remains unsolved. At the same time, the Khaidarkan mercury and Kadamjai antimony combines emit

Number of deposits where extraction of building stones and rocks is possible as of 1.01.2012.

Table 6

Types of raw materials	total fields, pcs.	Osh oblast		Jalal-Abad oblast	
		on the balance sheet	in operation.	on the balance sheet	in operation.
Natural building stones	66	3	1	2	2
Natural facing stones	65	1	1	2	1
Carbonate rocks	84	3	1	2	-
Gypsum	63	3	1	2	1
Clayey rocks	275	21	9	7	3
Clastic rocks	28	8	5	6	3
Raw materials for lightweight concrete aggregates	9	2	2	-	-
Active mineral additives in cement	1	-	-	1	-
Wollastonite	1	-	-	1	-
Bottom line:	592	41	20	23	10

8,000 tonnes of sulphur into the atmosphere each year in the form of gases and dust.

Sulphur dioxide is known to be easily captured by appropriate selective absorbents. Further, by chemical transformations, it can be converted into elemental sulphur or sulphuric acid. However, there is no technology in the region to capture sulphur dioxide gas from other gases and convert it into sulphur and sulphuric acid under industrial conditions.

There are also gaps in the utilisation of traditional mercury and antimony deposits. For example, during the years of Soviet rule, the existing mining enterprises were sufficiently supplied with explored reserves of raw materials. However, due to the changed ratio of prices for the products of mining enterprises, on the one hand, and for materials and equipment necessary for production, in particular, the sharp rise in energy prices, on the other hand, it became unprofitable to exploit the reserves approved according to previously developed specifications.

First of all, in order to develop existing mining enterprises, previously explored mineral reserves should be reevaluated. Deposits that are currently unprofitable to develop should be classified as off-balance sheet or held in reserve until market conditions change.

In the process of long-term prospecting in the south of the republic, a large number of deposits of non-ferrous and rare noble metals were identified, but they remained undiscovered and were not involved in development. With the development of business plans and proper organisation of production, it would be possible to create small enterprises for ore mining and processing.

The adopted Law of the Kyrgyz Republic "On Subsoil" gives great rights on protection and use of subsoil to management bodies. The State Agency for Geology and Mineral Resources of the Kyrgyz Republic has approved the "List of commonly occurring minerals" and the "List of deposits of local importance". The right to regulate the use of deposits included in the Lists is granted to the district administrations. The Lists include almost all deposits of construction materials, small deposits of coal, salts, gypsum, saw stone.

The main purpose of economic evaluation of mineral deposits is to determine the economic value and to select such parameters of deposit utilisation, which ensures the highest production efficiency both in the short and long term.

The economic evaluation of mineral deposits reflects the multistage process of prospecting, exploration, design and operation of deposits, mining, geological and technological factors. For example, for deposits of non-ferrous and rare metals, on the basis of the development of which mining, ore concentration and processing of concentrates at metallurgical plants are carried out, respectively, requires a complex technological scheme. At the same time, non-metallic construction materials, coal, gas and other types of mineral raw materials are produced using uncomplicated technological processes. In this regard, the complexity of economic evaluation varies significantly even for different mineral deposits.

In general, economic evaluation of mineral deposits is a multi-stage, multi-objective complex system.

Without touching upon the whole complex of issues related to the evaluation of deposits, we consider only those criteria and systems of indicators of economic evaluation of mineral deposits, which determine the efficiency of their exploitation. Thus, through the economic evaluation of deposits at the same time it is possible to determine the enomic indicators of the future mining enterprise in relation to the conditions of its production activity. Positive results of economic evaluation are widely used in the forecasting of the development of mining industries with the subsequent design and construction of industrial enterprises.

At the same time, it is important to note that various factors will influence the efficiency of the functioning of the planned enterprises based on the exploitation of mineral deposits. Thus, sufficient mineral reserves, favourable geographical and economic conditions, availability of water, fuel and energy, labour resources, as well as the level of development of social and industrial infrastructure will to a large extent predetermine the efficiency of operation of mining enterprises. Thus, it once again shows that the assessment of mineral deposits is reflected through the activities of the future mining enterprise.

The economic evaluation of mineral deposits uses a system of indicators that determine the economic efficiency and expediency of their involvement in the national economic turnover. Such indicators as cost, profit, profitability, specific capital investments, payback period of capital investments and other evaluation indicators are widely used in practice.

In the economic evaluation of deposits, the cost price, profitability, and profit fulfil the function of increasing their exploitation.

When evaluating different variants of non-ferrous metal deposits, in most cases it is impossible to ensure the equality of production volumes by variants, i.e. the condition for the comparative efficiency indicator is the same species volume of extracted components. But in this case, the range and volume of output for each of the compared options depend on natural factors, in particular, on the size of category reserves, differences in the composition and quality of raw materials. This is especially characteristic of deposits of complex raw materials. At the same capacity of the enterprise on extraction of raw materials volumes of output products can be different. In this connection, it is not a prerequisite for comparable variants at different deposits to be equal in terms of the volume of products obtained from them.

Differential mining rent is used as an estimator in the economic evaluation of mineral deposits.

Most economists believe that the category of differential rent is the most convenient as a basis for valuation of natural resources.

As a result of this discussion, a very valuable result was the recognition of the validity of the concept of differential rent.

In the mining industry, natural conditions participate in the formation of socially necessary costs. They create additional, so-called natural labour productivity, the level of which is determined by the quality, mining, geological and economic-geographical conditions of mineral deposits. Differences in natural labour productivity are stable, because the possibilities of involving relatively better types of mineral resources into exploitation are insufficient due to limitations and non-reproducibility. Therefore, higher natural productivity achieved by utilising relatively better sources of mineral resources cannot be the general norm.

The growing demand for mineral resources forces society to exploit mineral deposits with different natural conditions, which leads to sharp differences in individual production costs. As a result, differential rent is generated when exploiting the best deposits.

Consequently, differential mining rent is formed as the difference between socially necessary labour inputs determined by natural conditions of production and individual inputs.

Differential mining rent is the main part of the net income arising from the development of mineral deposits. In this regard, it acts as the main indicator of economic evaluation of deposits.

Wide involvement in exploitation of unique or located in the most favourable economic-geographical and mining-geological conditions of mineral deposits with the use of the latest achievements of science and technology, contributes to increasing the efficiency of the mining industry, expressed in reducing the cost of production of minerals, increasing the profitability of production, increasing the received profit.

It is safe to say that the further development of productive forces of the Kyrgyz Republic is connected with its regional peculiarities and rich natural resources. This refers primarily to the mountainous conditions of the territory of the Republic, where the main types of natural resources are located: mineral and raw materials, water and energy resources, resort, natural recreational resources, etc. The future economic potential of the Republic is inextricably linked to natural conditions and resources. Consequently, the future economic potential of the republic is inseparably connected with natural conditions and resources. Therefore, the issue of economic rational, ecologically clean and safe development of mineral and raw material resources in mountainous areas and the placement of mining enterprises, taking into account the changing demographic situation should become the objects of deep scientific research of economic and technical sciences.

Thus, the South of Kyrgyzstan has diverse and vast mineral resources. The task is to use them effectively, taking into account market relations. For this purpose, it is necessary, on the one hand, to conduct a unified scientific and technical policy on the part of the republican management bodies, on the other hand, to develop regional programmes, which would clearly indicate the organisational and economic mechanisms of rational, integrated and ecological development of mineral resources. This, in turn, requires the development of methods (basis) for the development of the mineral and raw materials complex of the region within the framework of the strategic goals of the republic.

This is how some aspects of natural conditions and resources look like for choosing the most acceptable options in further economic activity in the conditions of transition to market relations.

The assessment of natural resources contributes to finding practical ways of joint development of the biosphere and mankind, as knowledge of the factors of natural conditions will make it possible to find the best options for adapting the socio-economic system to the changes occurring in the biosphere and make the right national economic decisions.

Natural conditions affect social development in different ways at different stages, at different levels of productive forces.

Some zones, where walnut forests grow, are unique in the world in terms of combination of soil and climatic factors. Significant areas of natural pastures and hayfields ensure the development of animal husbandry.

The distribution of precipitation during the year is uneven. The greatest amount of precipitation falls in autumn-winter and spring periods. In July-September precipitation is insignificant - from 3 to 10mm, in some places 1525mm. The greatest amount of precipitation per year falls in the mountainous zone of Suzak, Aksy, Uzgen, Kara-Kulja districts - from 600 to 6900 mm, in other districts - from 300 to 520 mm.

The orientation of mountain ranges plays a major role in the distribution of precipitation in Southern Kyrgyzstan. The main sources of precipitation are western and south-western air masses, which cool down and condense moisture as they invade the region and approach the mountains.

On the slopes of mountains there is a difference in annual precipitation amounts depending on their absolute height and location, which have important ecological and economic values when making economic decisions. Thus, if at an altitude of 600-1000m the annual precipitation is up to 222mm, at an altitude of 2000-3000m it increases up to 6000mm. The amount of atmospheric precipitation increases from west to east. The most humidified areas are

Climatic conditions are largely determined by the peculiarities of geographical position and relief. Located in the subtropical latitudinal zone, inside a vast continent, it is remote from the paths of transport of humid air masses.

The climate is influenced by mountain ranges that enclose the territory of the region from the north, east and south and protect it from the direct impact of cold polar air masses coming from the north-east. The proximity of the Kara-Kum and Kyzyl-Kum deserts also affects the climate.

The belt extending to an altitude of 1100m is characterised by a semi-desert climate with moderately warm winters (average January temperature 3-40 frost) and hot, dry summers (average July temperature 24-270). This zone is most favourable for cultivation of cotton, vegetables and fruit crops, subject to irrigation.

In the belt located at an altitude of 1200-2000 m, the climate is moderately warm, winter is cold. The average temperature in January on the southern slopes of the

Chat-Kala ridge is 4-60, on the northern slopes of the Turkestan ridge - 70, on the slopes of the Fergana ridge - 30 frost. July temperatures are 15-200 degrees centigrade.

In the belt covering altitudes from 2000 TO 3000 m, the climate is temperate with long cold winters and cool summers, with temperatures up to 100 frost in January and 10-110 heat in July. In the belt above 3000 m, the climate is severe. There is no frost-free day. Snow falls on ridges and passes. Yak farming and horse breeding are developed in this zone.

The prevailing winds of these belts are winds from mountain slopes directed towards the axis of the Fergana Valley.

In the Fergana Valley, apart from the influence of the general atmospheric circulation on the climate, there is its own, internal system of air mass movement. Cold air, getting into the Fergana Valley, quickly warms up, transforms and in the plain areas does not significantly change the weather, but in the mountains the air temperature drops significantly, thunderstorms occur and precipitation in the form of rain and snow, and in May and June there is often hail and short-term heavy rain, which causes significant damage to cotton and tobacco crops.

Winter in the zone of irrigated agriculture is moderately cold. In the mountain zone it is usually established in early December. Its duration is 90-115 days, thaws are frequent. This factor is taken into account by farmers for making important agrotechnical decisions.

Spring comes early, but there are frequent returns of cold weather. March is characterised by the transition to positive mean monthly air temperatures. The second ten-day period of the month is characterised by stable transition of average daily air temperatures over +50. This determines the beginning of mass sowing of early spring crops and resumption of vegetation. At the end of March in the valley zone, the average monthly air temperature crosses +100, which significantly reduces the probability of frosts, allowing to start sowing of heat-loving crops. The maximum air temperature in March can rise to 20-270 degrees centigrade.

Summers are long and hot. The average monthly temperature ranges from 20-280, the maximum temperature in the valley-predmont zone rises to 35-430, in the mountain zone - to 30-350. July is the hottest month of the year. The duration of sunshine in summer is 100-120 hours per decade, and the maximum number can reach 120-135 hours.

Autumn is warm and dry. There is a sharp drop in air temperature from September to October. The average daily temperature in October ranges from 8-100 in the mountainous zone and 11-140 in the valley-piedmont areas. The average date of onset of the first autumn frosts in the foothill and mountain zones is 12-26 October, in the valley zone - 4-7 November. The intensity of frosts in October in the valley part can be up to 5-80, and in the mountain zone up to 10-200 below zero. The maximum air temperature in some years reaches 28-330. Proceeding from this, it is necessary to organise the work of the national economy in winter time, which requires significant energy resources and financial expenditures.

The creation of normal living conditions in the cold season depends on many economic, environmental and social factors.

2. State of land, water and mineral resources utilisation

Among the most important natural resources are land resources, whose place among other natural resources is determined by the functional characteristics of land. Land is the initial material basis for the well-being of the members of society, the spatial basis for the location of productive forces and human settlement, the foundation for the normal course of production processes, all factors of eco-comical growth - labour, material and technical, natural and entrepreneurial.

To characterise land resources as means of production, regional and territorial aspects are important. Table 7 gives a characteristic of land resources of the Kyrgyz Republic as of 1 January 1996.

Land resources located in different regions may differ significantly in their productivity due to natural features, soil, climate, topography, water availability, etc.

This difference is the basis for the emergence of additional income, differential rent on the best lands, which affects the results of economic activity in agriculture. This can also serve as a basis for regulating relations with non-agricultural land users.

Land resources are spatially limited and cannot be increased when free land is exhausted, unlike the possible expansion of production of other means of production. In addition, land plots are rigidly tied to a certain region, they are characterised by a permanent location, which causes the problem of creating special means of production for land use in a given region, development of the infrastructure system for interregional links.

The analysis of land use has shown that it is still low-productive in the republic. The main crops in the South of the republic are cereals (wheat and barley), corn, fodder crops (corn for silage and alfalfa), cotton, tobacco. The main problems of land productivity are related to soil erosion, salinisation of land irrigated by wrong methods.

Improved land productivity in a broad sense will depend on stable economic cultivation practices and good land-use management. The following types of fertility can be distinguished: natural, artificial and economic. Natural fertility is the result of millennia of geological. climatic, soil-forming processes. From the natural fertility, the availability of nutrients, moisture in the soil, their availability to agricultural plants largely depends on the output, and the realisation of natural fertility of the soil from the man himself, the level of agro-culture, the development of productive forces. The use of these factors makes it possible to significantly increase the original, natural fertility of the land. As for artificial fertility, it is created as a result of anthropogenic impacts. This constituent part of fertility requires additional means of labour, because "... part of the fertility of cultivated land is an artificial product, which owes its formation to culture and investment of capital".[1] The process of irrigation is also in the same line.

Irrigation is an important factor of artificial fertility. The analysis has shown that in the South, as well as in the republic as a whole, the irrigation network availability is low and does not exceed 41% (Table 8).

The combination of natural and artificial fertility forms economic fertility, which reflects the available capacity of the land to produce biomass. Quantitatively, economic fertility is expressed in the production of agricultural output per unit area, yield.

Characteristics of irrigated lands as of 1.01.15

Table 8 thousand hectares

Areas	Suitable for irrigation total	With irrigation network	Irrigated	Type of irrigation system		
				engineering	semi-engineered	non-engineering
South of Kyrgyzstan	724	310	296	135	72	90
Naryn	395	197	127	12	30	84
Issyk-Kul	296	185	163	29	84	50
Talas	354	154	127	5	48	74
Chuyskaya	477	380	324	221	92	11
By republic	2246	1226	1037	402	326	309

Four groups of measures to improve the use of land resources can be identified as part of a set of measures to green agriculture:[2]

1) improvement of the land fund within the framework of agriculture (erosion control, organic fertilisers, land reclamation, biological means);

2) organic influences of non-agricultural industries, sharp reduction of land withdrawal from agricultural turnover (construction of hydroelectric power plants, cities, mining industry, etc.), compensation of their losses by non-agricultural users, reduction of land resources;

3) search for and utilisation of vacant land suitable for agrarian purposes;

4) rationalisation of land fund use in terms of the final results of agricultural production.

At the same time, the analysis of the use of land resources has shown that the Republic, especially in the South, is not doing enough to combat erosion, deterioration of soil fertility, etc.

Cotton growing areas are the most frequently affected by pesticide contamination. Tobacco, one of the crops that requires large amounts of pesticides. However, the problem with pesticide use is not the total amount of pesticides per hectare, but rather the inefficiency of water supply and soil erosion, which leads to leaching of chemicals from fields and contamination of water resources.

The next important problem is water resources management. The Kyrgyz Republic has huge resources of underground and surface water. Significant reserves of water resources are available in rivers, eternal glaciers and snow massifs. Rivers at a certain stage of society movement began to have a great influence on the development of productive forces. Characteristically, the basins of large rivers played a major role in the development of the economy.

The distribution of rivers, lakes, glaciers and groundwater over the territory and the specifics of their regime are determined mainly by climatic factors, which in turn are strongly influenced by the factors of cellular relief differentiation. High mountains, being a barrier on the way of carrying air masses, have an extensive network of rivers, glaciers and lakes.

The plains adjacent to the Fergana Valley and low intermountain depressions are cut through by rare "transit" rivers, which here lose their water for evaporation, irrigation, filtration and often end up with "blind" estuaries. Thus, hydrological areas of water flow formation and dispersion are distinguished.[1]

There are about 600 rivers, creeks and streams with a length of 10 kilometres in the region. All of them belong to the basins of 3 major rivers of Central Asia: the Syr Darya, the Amu Darya and the Tarim. Amu Darya and Tarim. The largest river in terms of water content in southern Kyrgyzstan is the Naryn River, which crosses the territory of Jalal-Abad oblast in its lower reaches, and the Kara-Darya River in terms of catchment area within the oblast. The Amu Darya and Tarim basins occupy only a small part of the region. The list of the most water-bearing rivers of economic importance is given in the table below.

The most water-bearing rivers of Southern Kyrgyzstan[2]

Table 10

Name of rivers	Catchment area	Annual runoff, mln m^3	Average annual consumption, cubic metres[3]
Karadarya	30100	3847,6	121

	5520	2712,0	122
Chatkal (within Kyrgyzstan)	4420	1431,7	45,4
Thar	3510	1327,6	42,1
Soh	1675	1315,0	25
	2470	1238,6	39,6
Kyzyl-Suu (within Kyr-Gyzst.)	2620	1191,1	34,8
Kara-Suu (right)	4130	905,1	29,6
Jazzes	3750	826,2	26,2
Kara-Ungyur (Tentek Sai)	2220	693,8	21,1
	2540	674,9	21,4
Kurshab (within Kyrgyzstan)	1200	671,7	21,3
Isfayram-Sai	1370	570,8	18,1
Ak-Buura	3240	450,9	14,7
Kara-Kulja	2150	346,9	11,0
Kögart	1780	340,6	10,8
Isfara	1960	327,6	10,4

The rivers of the south-western slope of the Fergana Range pose a great environmental hazard. These rivers include: Kara-Kulja, Dzhazy, Kyogart, Kara-Ungyur (Tentek-Sai) and Maili-Suu. They are characterised by spring-summer floods formed as a result of melting of seasonal snows, which annually cause great material damage to the national economy. The catchment areas of this group of rivers are located relatively low (up to 4000 m above sea level). Along the Alai, Turkestan, Fergana and Chatkal ridges, there is a strip of relatively low mountains, passing into the foothill plains of the Fergana Valley. There are widespread beds of intermittent drainages fed by melting seasonal snow or rainfall, which often cause the formation of mudflows.

A characteristic feature of the hydrogeographical network of the plains is the abundance of artificial water structures: canals, ditches, reservoirs, etc. The main feature of the hydrogeographical network of the plains is the abundance of artificial water structures.

There are more than 100 lakes in Southern Kyrgyzstan. Most of them are small, often of rubble origin, when moraines and mountain debris blocking river valleys serve as dams. Such lakes in the areas of ancient and modern oleation were formed in karas and glacial sandstone fields. There are also karst lakes in places where easily soluble rocks are deposited. Typical water reservoirs of oval-

The large lakes Sary-Chelek in the basin of the Pachata River, the left part of the

Kara-Suu River at an altitude of 1858 m above sea level, and Kulun in the basin of the Tar River at an altitude of 856 m (see Table 11). Kara-Suu at an altitude of 1858 m above sea level and Kulun in the basin of the Tar River at an altitude of 856 m (see Table 11). There are 16 more lakes with a water mirror area of more than 0.1 km. The rest of the lakes are small, with water surface area from 0.09 to 0.01 km.

There are no saline lakes of natural origin. Table 11 shows the characteristics of the largest lakes in the South of Kyrgyzstan.

Major lakes in the South of Kyrgyzstan

Table 11

Name	Water mirror area, sq.km.	Altitude above sea level, m	A river flowing out of a lake
Sary-Chelek	4,52	1858,6	Sary-Chelek
Kulun	3,25	2856,0	Kulun
Kök-Ala	0,84	2606,3	Gawa-Sai
Ai-Kyol	0,82	2637,0	-
Kutman-Köl	0,54	2836,0	Kerey
Kulun-Maly	0,45	2505,2	Kulun
Kara-Suu	0,4	1870,0	Kara-Suu

In Osh and Jalal-Abad oblasts there are more than 2,033 glaciers with a total area of 1,852.2 km. They are distributed mainly on the northern slopes of the Chonalai, Alai, Turkestan and Kichik-Alai ranges. Separate centres of small forms of modern glaciation are found on the slopes of the Fergana and Chatkal ranges. (see Table 12)

Distribution of glaciations by ranges[1] Table 12

Ridges	Number of glaciers	Area of glaciers, sq.m.
Chatkal	181	61,8
Fergana	270	88,2
Kichikalay	77	87,7
Alay	1034	676,4
Turkestan	246	244,8
Chon-Alai	215	693,3
TOTAL	2023	1852,2

Why are these data cited? The ecological explanation is as follows. An increase in the concentration of carbon dioxide in the atmosphere can cause global changes in the Earth's climate due to the so-called "greenhouse effect" of the planet's air envelope. The essence of this phenomenon is that a layer of air enriched with carbon dioxide transmits solar radiation well and at the same time retains the Earth's long-wave thermal radiation. The sunlight reflected by the Earth's surface in its infrared region is absorbed in the atmosphere, leading to an increase in its temperature. It has been calculated that an increase in the content of carbon dioxide in the atmosphere should already at present increase the average global air temperature near the Earth's surface by 0.3-0.40C. Such an insignificant, at first glance, increase in temperature may, however, cause significant changes in the existing ecological systems. It should be remembered that water at zero degrees is in a triple aggregate state: vapour, liquid, ice. A further change in the Earth's albedo, providing an increase in air temperature by 2^0 C, will lead to melting of ice, rise in surface water levels, and catastrophic flooding of the Earth's surface.

However, it is worth remembering that nature has self-regulation and self-restoration properties. And the given data is necessary for taking appropriate engineering measures.

In economic terms, glaciers play a regulating role in the hydrological regime of many rivers - they melt in summer, when fields need irrigation the most and a significant part of river water is used for irrigation.

Powerful mountain ranges occupying most of the territory are the richest accumulators of moisture (snow, glaciers) and give rise to the main rivers. The Naryn River is characterised by high water content and is of particular importance for hydropower generation.

The rivers originating from the slopes of the Fergana Range play a huge role for irrigated agriculture. These are mainly Kara-Darya, Maili-Suu, Kyogart, Kara-Ungyur, Ak-Buura. These rivers are fed by melt water and rain in spring and by eternal snow and glaciers in summer. The uneven distribution of water flows by month has necessitated the construction of reservoirs on these rivers in order to conserve water for irrigation during the critically dry summer months. However, this on the other hand created environmental problems.

The great complexity of the geological structure of the territory of the region, the diversity of types of rocks composing it and the duration of their formation, relief and climate, as well as human activity in terms of irrigation and water supply have determined the peculiarities of hydrogeological conditions and groundwater resources.

Within the mountain ranges, the conditions of groundwater formation differ from those in other areas. Groundwater is accumulated in fractured zones formed up to a depth of 100-150 m and in fractured rocks of large faults. This causes their extremely uneven distribution and relatively small resources per unit area. Groundwater of mountain ranges is fresh and is used for watering of pastures of distant cattle breeding and agricultural water supply by means of spring dripping and, less often, drilling of wells. Often groundwater resources available here are sufficient to solve water supply issues in mountainous areas. The productivity of individual wells does not exceed 2-3 l/sec. By chemical composition and degree of mineralisation they are very diverse. Dry residue varies within 1-10 g/l, and chemical type is sulphate-chloride-calcium. Groundwater of the foothill zone is also used for watering pastures of distant cattle breeding and water supply of villages.[1]

All this has a direct relation to water resources management, as uniform distribution over zones and territories is a guarantee of good crop yields and use of agrotechnical methods.

Groundwater is important in analysing water resources management. They are formed on the areas occupied by intermountain depressions. Groundwater is mainly fed by precipitation and river seepage, and only 3-5% by groundwater inflow from the mountains. As a rule, in most depressions, groundwater is of good drinking quality (mineralisation rarely exceeds 1 g/l), and the chemical type is hydrocarbonate-calcium. Intermountain depressions of the region are the most economically developed territories. The majority of settlements are concentrated within their boundaries, the intensive use of groundwater for various needs of the national economy is widespread: there are 13 depressions within the region, the groundwater of which is promising for use in the national economy. The total amount of water reserves in 13 deposits under the 24-hour operation mode is 1916.4 thousand cubic metres per day.

Nine self-spouting wells (with total flow rate of 17.4 l/sec) of Jalal-Abad and Karashorinsk thermal mineral water deposits are used for therapeutic purposes and industrial bottling.

Kyrgyzstan has the richest hydropower resources. According to experts' calculations, the potential hydropower resources of Kyrgyz rivers are estimated at about 142 billion kilowatt/hours. On the Naryn River and its tributaries alone, it was planned to build 16 HPPs with an annual output of 30 billion kilowatt/hours of electricity.

As studies show, the sources of hydropower resources in Kyrgyzstan are reliable, they are formed by eternal glaciers and snows of the mountains. This means that they carry with them the notion of inexhaustibility, in ecological terms electricity is the most reliable and clean, and does not harm the natural environment.

The Toktogul reservoir provided annual and multi-year regulation of the Naryn river flow, allowed additional irrigation of 400 thousand hectares of new lands in Uzbekistan and partially in Kazakhstan. If to calculate all profit received for the last 20 years since Toktogul HPP commissioning, as a result of growing cotton, rice and other agricultural crops on new lands and increased water supply of over 900 thousand ha of existing irrigated lands.

Thus, water resources management is largely related to the use of water in reservoirs. Meanwhile, there are some problems here. For example, the Toktogul hydroelectric complex and its reservoirs provide only electricity to the republic. The lands are not irrigated, as it is built below the agricultural lands of Kyrgyzstan. The Toktogul reservoir, built in the Ketmen-Tyubinsk valley, occupies 322 _ thousand hectares, including 12 thousand hectares previously classified as irrigated land.

Four settlements and the district centre, where about 30 thousand people lived, remained under water. These people were resettled to new places, in mountain gorges. In Soviet times, the turbulent Kara-Darya River was curbed. Its flow was regulated, otherwise there would be no guaranteed water supply for cotton plantations and sustainable yield of this crop in the Fergana Valley. It was decided to build the Kampyr-Ravat reservoir in Uzgen district (now called the Andijan reservoir) with a capacity of 1,750 million cubic metres of water.

At present, the Andijan reservoir has accumulated 1.2 times more water than it was designed for. As a consequence, the groundwater level has risen much higher in the nearest villages of Uzgen and Kara-Suu districts, it floods houses and destroys various buildings. During the growing season, water erodes the banks of the Kara-Darya River.

Toktogul, Papan reservoirs in the South of Kyrgyzstan are built in high 8-9 point seismic zone. This complex economic and ecological task requires large material and financial resources. The ecological danger lies in the unpredictability of severe consequences if many billions of cubic metres of water accumulated in these reservoirs, due to a possible natural cataclysm, spill over fields, villages and cities. Therefore, we need a material and technical base to maintain the dams of large irrigation facilities in a technically sound condition. And from the economic point of view, the republic is already suffering great damage. From the Toktogul reservoir during the growing season, an average of 70% of the annual discharge volume is withdrawn for the needs of neighbouring countries. This leads to reduced power generation in winter time. The annual damage is estimated at about 280-300 million KGS.

Critical water deficit is created in two provinces of Kyrgyzstan that have large water resources - Osh and Jalal-Abad provinces. These provinces, together with Naryn province, had a limit of 14.5 per cent of their own reserves during the Soviet period.

Therefore, a common strategy for the development of water and hydropower resources of the republic is necessary. For this purpose it is necessary to adopt interstate agreements: on justified and specified water limits from transboundary rivers on joint use of water and energy resources; on rules and tariffs of payment for water and electricity used in the national economy of the parties; on trade in natural resources and the order of mutual settlements.

Environmental problems also lack a legal basis. There are no Laws on the protection of surface and underground water, soil and air environment and on compensation for damage from their pollution. This is all the more necessary if we take into account the use of natural resources as drinking water. A number of cities and settlements in the South of the Republic often use polluted water as drinking water.

The main source of water pollution is spoilt distribution systems and contamination of surface water sources. Long-term observation, for example, has shown that the main source of Osh is the Ak-Buura River, where sudden and destructive flows are frequent. The river basin is home to a large number of livestock and agricultural farms, as well as many dachas without the necessary sanitation facilities.

The environmental situation is aggravated by the fact that there is no systematic assessment of the effectiveness of biological means of wastewater neutralisation, the collection capacity of which is approximately 70% of the water supply capacity.

According to the physiographic zones characteristic of the territory of the region, several types of desert vegetation (original in their life form prickly-pods), diverse steppes, high-grass, subalpine and alpine meadows (pistachios, almond trees, alichovniki, exochordniki and other shrubs), relict nut-fruit forests, fir, spruce, maple, birch, poplar-willow, juniper forests and others have formed. Screes and rocks are covered with ^formed plant groupings, and in high mountains - with cryophilic pads.

Wormwood-ephemeral deserts are used as fodder lands (yield from 3 to 8 c/ha). Among the deserts, especially on stony-debny soils, thorny-podded plants are found.

Steppes of the region are widely spread. The virgin areas are important fodder lands (yield from 5 to 18t/ha). Ferruginous, prangos, eremurus steppes are the place of medicinal raw materials procurement.

Tall grass meadows are located at an altitude of 1200-1400 m above sea level, and below these altitudes along river floodplains. They occupy forest glades, occur among shrubs and sparse forests. In terms of floristic composition, meadows are rich and diverse. These meadows are used as hayfields and pastures for cattle (yield up to 4 c/ha).

The forests of the region are diverse in floristic composition. The most valuable and interesting are unique nut and fruit forests. They are formed on the slopes of the Fergana and Chatkal ridges at an altitude of 1200-2200 metres above sea level. Here, along with walnut, apple, pear, currant and alcha grow. At the upper border walnut-fruit forests are replaced by spruce, spruce-fir, maple forests. These forests have a park character and are interspersed with meadows and shrubs. Juniper forests are widespread, especially on the Turkestan and Chatkal ridges. Here are the largest massifs of juniper forests in Kyrgyzstan. Juniper forms thin forests and sparse woodlands. Meadow and meadow-steppe species are expressed in the grass cover.

Poplar-willow forests are most abundant in river floodplains and rarely on river slopes, interspersed with various shrubs and meadow vegetation. Forests play a great role in the nature of the region, fulfilling water protection, anti-erosion, soil protection and sanitary-hygienic functions. Walnut and fruit forests are annually harvested for walnuts (up to 1000-1200 tonnes), apples (up to 5000 tonnes), alycha, tannins and other minerals.

Shrubs are widely spread. Pistachio and almond trees are found. Pistachio and almond fruits are used for food industry and medicine. There are widespread rose gardens (from rose hips of Kokandsky and flat-topped rose hips), barberry (from barberry), the fruits of which are used for preparation of vitamin raw materials. Shrubs of hawthorn, honeysuckle, tabylga, irga, and turmeric also have water protection, soil protection, and erosion control value.

Different environmental conditions associated with the vertical zonation of the region's territory have a significant impact on the species composition of the fauna. Mammals are represented by 65 species. Birds are represented by 22 species. Many wild animals, fish and birds are commercially important.

As a result, the region has sufficient resources of land, water, bioresources, etc. Increasing the level of their management is a task not only for the near future, but also for the distant future. In this regard, there is an urgent need to prepare. National strategy for the effective use, management and conservation of resources.

3.The state of the Kyrgyz economy in transition and assessment of state regulation of nature use

State regulation is important in the management of national natural resources. The economic and regulatory functions of the state are diverse. They in the use of natural resources are mainly related to ensuring appropriate rates of economic growth, ensuring the legal framework, protection and preservation of the environment, foreign economic activity and others.

The results of human impact on nature must be considered not only in the light of technological development and population growth, but also in relation to the social conditions in which they occur; different socio-economic systems are characterised by different human relationships with nature.

The economy of Kyrgyzstan before the perestroika times was based on the intra-union division of labour, it was originally based on disproportions, and therefore it was supported by subventions, which amounted in 1989 - 1991 (the last years before sovereignty) to 10...13% of the gross national income.

In the economy of Kyrgyzstan there are two unconnected complexes: branches of the agro-industrial complex (AIC) with their predominantly raw material orientation and machine-building industries, as well as mining and metallurgical plants, which are almost completely dependent on orders and supplies from outside the country. As a consequence, there is economic and social disintegration of cities and villages, enclave character of industrialisation, when industry is concentrated only in large industrial centres.

The main course of economic reforms in modern Kyrgyzstan is aimed at introducing market relations of the economy, which determine the price depending on the ratio of supply and demand and, vice versa, regulating the ratio of supply and demand depending on price fluctuations. It should be noted that market relations are still underdeveloped and imperfect.

Competition - the struggle of producers among themselves for a greater share of profits, for markets, for sources of raw materials and other factors did not work. As a result, the economy is doomed to stagnation regression.

Most importantly it fails to launch the marketing mechanism of free competition.

Kyrgyzstan, having become an independent state, entered the world community as a developing country in transition on the way to building a democratic state with a market economy. Despite the refusal of the state order, the planned economy does not think of outliving itself. Its principles, which include set parameters such as volume, order of work, orientation, regulation, projects, are still alive.

Thus, at present, with all the multi-structuralism of the Kyrgyz economy in the transition period from the planned system to market relations, it is distinguished by the presence of three main economic patterns: non-expired elements of the planned economy, mercantilism as an intermediate pattern and the market - the leading pattern.

What has been done during the years of economic reforms in Kyrgyzstan? The foundations of a market economy have been laid: institutions of private property, private enterprise and free entrepreneurial activity have been created.

In the system of the former USSR in other countries with planned economies in Kyrgyzstan, the main focus was on increasing the volume of a narrow range of production - industrial giants and quantitative indicators, and the prices of natural and capital resources were extremely underestimated. The Republic under this system was a raw materials appendage. Energy carriers and natural resources were used extremely irrationally due to their underestimated value. "The panacea for all production woes was additional capital investments.

In Kyrgyzstan, natural resources are the exclusive property of the state, and therefore it is proposed to call them "national". Under such conditions, liberalisation in this area of the economy should be gradual and selective. Since the market and planned economy will coexist for some time, individuals and companies will have a strong incentive to extract economic rents by moving goods or financial resources from cheap and controlled economies to liberalised and higher priced economies. The government, as a basic instinct of the state, will have to impose strict controls at both macro- and micro-economic levels on those activities that are still subject to the plan and impose stiff fines on violators.

The essence of management under socialism was determined by the whole system of socialist production relations and, above all, by systematicity as the universal and initial form of production relations of socialism.

The leading and guiding force in the system of management of the national economy under socialism was the Communist Party. The state acted as the main centre for the management of the socialist economy. The objective basis for the transformation of the state into the main centre for the management of social production was public ownership of the means of production.

The most important principles of socialist production management were: so-called democratic centralism, party leadership, the leading role of the state in production management, plannedness.

Liberalisation of the economy and tight monetary policy conducted in accordance with the programmes of economic transformation allowed Kyrgyzstan to achieve certain successes in key macroeconomic indicators. The implementation of the programmes made it possible to reduce inflation, strengthen the national currency, and initiate structural changes in the economy.

At the same time, the general economic condition of the Republic is influenced by the weakness of the state and its executive bodies in the formation of market relations and regulation of the economy due to limited instruments of influence. In the conditions of the transitional economy the conditions for regulation of the economy under the influence of market forces are not yet ensured and serious efforts on the part of the state are required to achieve the movement of economic processes in the right direction.

A sharp change in the political regime, radical restructuring of the economic system also requires the use of appropriate mechanisms and tools of state regulation of economic processes. There is a vital need to develop a comprehensive medium- and long-term government policy that would ensure balanced and sustainable growth of the economy.

CHAPTER 3

MAIN DIRECTIONS FOR IMPROVING NATURAL RESOURCE MANAGEMENT

1. The impact of improved natural resource management on improving public health and people's standard of living.

In the complex of problems related to the life and development of human society in the conditions of transition to a market economy, the problems of finding an optimal balance between its economic needs and the opportunities provided by the environment are of particular importance. This problem requires the adoption of conceptual solutions and practical measures.

The problem, first of all, is to find a positive impact of the process of nature management on the living standards of the population. This is understandable, as the degree of utilisation of existing reserves of natural mineral and raw material resources largely determines people's income security. In addition, more efficient and full use of natural resources contributes to the reduction of unemployment, more efficient distribution of resources, application of improved forms of labour and technology.

However, while attaching great importance to the role of natural resources in economic development, we should not forget about the conditions that ensure such development. The fact is that some countries such as Angola, Ethiopia, Brazil, Nicaragua, etc., having abundant natural resources, lag behind in economic development, while others such as Japan, Malaysia, South Korea and others, having no resources, are making tremendous progress. The reason lies in the different material and socio-cultural conditions in which this or that country is located and the state of scientific and technological development.

Kyrgyzstan is a developing industrialised country. Despite the availability of certain natural resources, the republic is currently experiencing an acute shortage of investment for the acquisition and creation of goods, production facilities, machinery and equipment, social infrastructure, etc. There is no doubt that the increase in the stock of labour can increase its productivity and raise the standard of living, since there is a close relationship between labour productivity and real income per employed person.

Unfortunately, the situation in the Kyrgyz Republic is such that the standard of living of the population is low and poverty prevails, which cannot be precisely defined. But in a broad sense we can say that a family lives in poverty when its basic needs exceed the means available to meet them. In the future, the problems of poverty may worsen as the transition period has seen some aggravation of poverty, at the moment half of the population is categorised as poor, with a third of it living in extreme poverty. We bring these arguments to show the correlation between the living standards of the population and the use of natural resources to create appropriate prerequisites and conditions for people's livelihoods. It is no secret that it is the rational use of land, water, mineral and raw material resources that is ultimately the basis for people's livelihoods.

To improve land use, it is imperative to introduce soil conservation technology and improve irrigation practices.

Improving the storage and use of agricultural chemicals is also a priority.

A certain result in the increase of agricultural production is the improvement of pastures. Unfortunately, for many years in the republic there has been a constant work on increasing the number of animals without taking into account economic conditions or the state of the environment. As a result, today we have pasture resources, more than 50 per cent of which are degraded due to grazing of large numbers of animals. In addition, both before and now, in the absence of market status of pastures, many people consider them as a form of preservation and accumulation of wealth. In these conditions, uncontrolled grazing and trampling of pastures and other negative manifestations are possible.

Therefore, in the coming period the republic urgently needs to develop a system of pasture use with the definition of scientifically-based norms of herd structure for each hectare of pastures, depending on their productivity.

One of the main directions for improving the use of land resources is the successful implementation of land and agrarian reform in the republic.

The soil cover of the earth is a renewable natural resource, but its restoration by natural means requires hundreds of years (restoration of a soil layer 2.5 cm thick requires 300...1000 years, and for the whole arable layer with a thickness of 18 cm - 5...7 thousand years), therefore it is necessary to be extremely careful in land use. The right approach to land use, its global saving and protection allow not only to maintain its biological potential and productivity, but also to multiply them significantly. Decrease in fertility is largely associated with subjective factors, primarily with weak implementation, achievement of scientific and technological progress, insufficient use of the possibilities of artificially created means of production in agriculture. At the same time, in a certain period, the belief that machinery, fertilisers, pesticides can infinitely increase yields led to stagnation in the development of agricultural culture, ignoring the natural features of the land. Therefore, in the coming period it is necessary to focus not only on reproduction (simple and expanded), economic fertility, but also on the creation of certain conditions for neutralisation of subjective factors.

The first chapter noted the need for land valuation based on compensation payments to compensate for damage to agricultural production when productive land is allocated for non-agricultural purposes (for industrial, infrastructure, urban development) and separately for urban land valuation. Compensation payments, which depend on many factors, should be used in developing new land for agriculture or improving the fertility of existing land. As for urban land, the proceeds from its sale are spent on engineering development of land, creation of infrastructure, etc. It is understandable that in the long term, special measures need to be developed in these two areas in order to address the serious gaps mentioned above.

A reliable direction for improving human life is undoubtedly the provision of freedom of choice of economic activity, increasing motivation and incentive to work, efficient farming, etc. The role of land and agrarian reforms is crucial. As noted, the ongoing land and agrarian reforms play a decisive role in this endeavour. However, in the future it is necessary to legislate many aspects of land use. In addition, it is necessary to constantly work on increasing the adaptive potential of cultivated plants with simultaneous development of scientific systems of designing ecologically sustainable and energy-saving agro-biocenoses with maximum consideration of environmental factors and creation of new plants based on the use of biological and breeding potential of cultivated plants.

There are certain reserves in the utilisation of forest resources to increase the products for the needs of the population. First of all, they play an important role in soil deregulation and protection. Sheep and cattle grazing on the forestry lands provide additional food products. In addition, we should not forget the essential role of fruits and nuts (such as walnuts, almonds, pistachio, apples, pears, apricots, alycha) in replenishing the food stock. Another valuable forest by-product is honey, especially in Jalal-Abad oblast.

Another source of improving life through better utilisation of natural resources is the development of tourism, especially ecological tourism. The south of the republic has great opportunities in creating tourist routes by organising hiking, mountain climbing, horseback riding, skiing. The presence in the region of blue lakes (Sary-Chelek, Kara-Suu) and crystal clear rivers, as well as multi-coloured valleys and deserts create a unique beauty. Wild animals are unique and diverse, and the region is rich in historical events. Osh is one of the oldest cities in Central Asia and celebrates its 3000th anniversary in 2000.

Trophy hunting can be organised in the South of the Republic to a certain extent. It is possible to organise tourist routes in cooperation with neighbouring countries (e.g. Uzbekistan), visiting such cities as Samarkand, Bukhara, Tashkent, Andijan. For this, much work needs to be done to create the infrastructure, facilities and services that travellers require. There are boundless opportunities for the industry of the republic in improving the utilisation of natural resources. Therefore, a strong industrial policy is needed to utilise natural resources for the needs of the population. For this purpose, first of all, it is necessary to intensively develop raw material resources, which are partly not yet fully explored. For example, the development of over 1200 deposits and ore occurrences in the region. These are deposits of gold, silver, mercury, tin, coal, tungsten, antimony, etc., and there are great prospects for the development of the construction materials industry.

In the future, the stone processing industry could be developed in the region. Naturally, there are great opportunities in the South for further development of the energy sector and improvement of hydropower resources, which are renewable.

Following the outline of the main directions for improving the utilisation of natural resources in order to raise living standards, it is necessary to take appropriate measures to improve the health of the population. The main health problems are related to poor child nutrition, hepatitis A (especially in the South), tuberculosis, maternal and infant morbidity and mortality.

It is time to develop a national plan for environmental health care, which in a market economy, with the emergence of multiple forms of ownership, can be developed both through public support and funding and through private and collective medical practice.

The next step in improving public health through better utilisation of natural resources is to eliminate the influence of water containing bacteria and viruses. In this case, the situation is particularly unfavourable in the South of the Republic, where, due to the large amount of irrigated land, canals and other open water sources are contaminated by irrigation water runoff as well as by freely grazing farm animals.

The measures taken here should be primarily aimed at improving the sewerage system, sanitation services for the population, etc.

As mentioned above, the content of mercury and antimony vapour in the atmospheric air of Khaidar-Kana and Kadamjai has a very harmful effect on human health. However, in these urban-type settlements, there are no sanitary or protective zones separating residential areas from the mines. In these cases, a monitoring programme that would identify practical measures to reduce harmful effects on the health of people living in the area is extremely important.

Timely collection of information on the morbidity of the population and its processing could make some positive contribution to the improvement of public health, so that specific measures could be taken by the relevant authorities to reduce harmful substances on morbidity and mortality, ensure the quality of drinking water in rural areas, and eliminate the effects of pesticides in crop cultivation. Such information should be available to both the services of the Ministries of Environment and Health and economic services, such as water supply, SES and others, in order to coordinate and unite efforts to control the morbidity of the population.

Among the factors affecting human health are some phenomena polluting the atmosphere, in particular radioactivity of some industrial wastes. In the past, strategic raw materials were extracted from Kyrgyzstan, including from the southern regions. Unfortunately, when developing deposits of uranium and other rare earth substances, no attention was paid to ecology. A serious problem in this regard is the burial of dumps from uranium deposits in Maili-Suu. This problem, handed down since the times of the USSR, poses a certain threat not only to the inhabitants of Kyrgyzstan, but also affects the interests of the population of the neighbouring state - Uzbekistan. Therefore, international coordination in solving this issue is required.

Not only in rural areas, but also in urban areas, waste collection and neutralisation systems are inefficient or not working. Often solid waste is not removed in time, and about 90 per cent of solid waste dumps in the cities of Osh, Jalal-Abad, Sulukta, Kyzyl-Kiya and others do not meet sanitary standards. It is clear that such phenomena should not be allowed. Local and city households should seriously take care of creating a landfill for storage and burial of toxic and other industrial wastes, including oil, other organic substances, timely repair of sewage systems.

To summarise the above, health care interventions should have a real impact on improving health care in such forms as reducing disease, reducing child morbidity and mortality or minimising risk factors so that any interventions taken serve human health to the maximum extent possible.

2. Measures on environmental protection in the system of nature management

Environmental protection is of great importance in the system of natural resources management, the essence of which comes down to the fact that by means of reasonable regulation of society's impact on nature it is possible to ensure the preservation, restoration and, where necessary, improvement of favourable conditions for human life, development of production and culture.

Inadequate and untimely environmental protection measures can lead to negative consequences. Appearance of Papan and Andijan reservoirs along with other factors in Fergana valley in due time negatively affected the condition of some agricultural crops. Thus, as a result of changes in climate-forming factors under their influence, the effective temperature of solar energy necessary for cotton opening is not accumulated. The volume of gross cotton cultivation in Kyrgyzstan has been reduced to 70,000 tonnes from 210,000 tonnes by now. As a consequence, the Osh Cotton Production Association is not supplied with raw materials.

These and other artificial seas stabilised groundwater rise. As a result, the Western microdistrict, Tyuleiken village and other settlements of Osh city, part of the territory of Kara-Suu district are flooded. Some areas have been turned into swamps.

With the commissioning of the Toktogul reservoir, more than 32 thousand hectares of agricultural land were taken out of circulation.

By interfering with natural processes with the help of modern technology, man often violates the regularities of their course, his industrial activity causes undesirable changes in nature for himself. In the South alone, over the last 10-15 years, forest areas have shrunk by more than 20 per cent and many species of animals have disappeared.

In Batken and Kara-Suu districts, the areas of eroded lands are increasing, and anthropogenic pollution of water and air has started everywhere.

Environmental protection problems are largely related to the operation of industrial enterprises and farms in the agricultural sector. In each particular case at an industrial enterprise, plants or factories, in industry or in agriculture, in transport or in services - a certain amount of natural resources is functioning, as a result of the use of which production waste, emissions, etc. is formed, which, getting into the air, water, pollute the environment.

One of the ways to positively influence the work of these enterprises is the strict implementation of legislative provisions in the field of environmental protection. As an example, the Law of the Kyrgyz Republic "On Nature Protection", adopted on 17.04.1991, where in the section "On Sanitary-Epidemic Welfare of the Population", it is stated about the right of citizens to a favourable environment, which could not have a negative impact on the health of the present and future generation. Unfortunately, today, in many places and cases, this provision of the Law remains a declaration.

Thus, as of 1.01.1996 in Osh oblast, out of 116 industrial enterprises, 60 do not have sanitary protection zones, including Osh city - out of 33 in 24, Kyzyl-Kie - out of 14 in 4, Sulukt - out of 24 in 3, Kara-Suu - out of 13 in 13, Uzgen - out of 22 in 12. Kadamjai district - out of 7 in 4[1] .

According to the same organisation in 1996, 301 out of 2186 samples taken for air pollution did not meet the MPC. The most polluted air is still the air of Osh, Kyzyl-Kiya, Kadamjai, Aidarke-n.

Pollution in open water bodies is still high. Out of 196 samples, 22 samples do not meet the norms in terms of microbiological indicators. In 16 samples pathogens of infectious diseases were detected. These are sources of economic water supply to the population.

The main polluters of water bodies and soil are still dairy farms and livestock complexes of farms. At present in Osh oblast under different ownership there are 1 million 700 thousand cows and steers, about 7000 horses, more than 3000 pigs, and almost half a million poultry. Because of this, in 1995 in Osh oblast there was a sharp increase in blood and blood-borne diseases, pneumonia, bronchitis, kidney infections. Here it is necessary to intensify the work of rural service centres and veterinary stations, the number of which is now 130, but this is not enough for the region.

In 1995, 54.4 tonnes of pesticides, 48 tonnes of deroliquites and other chemicals were used for chemical pest control in Osh oblast. However, due to their unsatisfactory storage, the environment was greatly damaged. In total, there are 33 warehouses in the oblast for storage of pesticides, 12 of them do not meet sanitary requirements. Out of 14 dressing sites 5 are completely unsuitable[2] .

In 1996, the contamination of food and agricultural products according to the results of laboratory tests of the Osh Regional Sanitary and Epidemiological Station (oblSES) was 15.5 per cent. It increased by 2.4 per cent compared to 1995. In total, the central laboratory of the oblSES conducted 1552 tests of foodstuffs and environmental objects for contamination with residual quantities of pesticides in 1995. In 79 tests contamination with organochlorine compounds was detected, in 44 - with organophosphorus compounds, in 9 - with pitetroids and 32 - with pesticides.
other pesticides of different chemical classes. Gosannadzor bodies of the region took measures to prevent pollution of environmental objects. In 1996, 38 fines were imposed for poor protection of water bodies, 35 materials were submitted to administrative commissions. Operation of 15 objects was suspended, 32 fines were imposed for air pollution and 126 fines were imposed for soil pollution.

Rational use of water resources and preserving the purity of natural water bodies is one of the most important aspects of environmental protection.

Here we would like to give some ways to improve their use. As priority economic measures to improve the use of water resources, in our opinion, the national economy should face the tasks of all-round economy of water resources, protection of river basins, lakes and other sources from pollution, clogging and depletion, reduction of specific water consumption per unit of production. All of the above shows the urgency and acuteness of the problem of water supply and water use at the present stage and the need to increase the effectiveness of water-saving and water-protection measures aimed at protecting water bodies from pollution and depletion, the introduction of low-waste technological processes, as well as the development of new methods and facilities for the treatment of industrial and municipal wastewater. It is necessary to take measures to ensure the rational use and protection of water resources from pollution and depletion, to increase the capacity of water recycling and water reuse systems, and to develop and introduce drainless water use systems at enterprises. The activity of water management units, the main purpose of which is modern provision of consumers with water of required quantity and quality, should play a decisive role in this.

The material base of the "water management" sector includes: hydraulic structures (dams, dikes, canals), i.e. all fixed assets that serve the purpose of rational use of water resources at the scale of water source participation; infrastructure facilities that allow using water for the production process (pumps, pipelines, water cooling systems); fixed assets for water protection purposes (treatment facilities).

In our opinion, large expenditures necessary to maintain water resources in good condition objectively require state regulation, as many objects of the water management system are objects of inter-farm, all-regional and all-republican purposes.

The rational use and protection of water resources should be based on the water balance, which contains resource and expenditure parts. Indicators of rational water use are: the ratio of wastewater disposal to the volume of fresh water received; water use multiplicity - the ratio of gross consumption to the volume of fresh water consumption; the number of enterprises stopping the discharge of untreated and untreated wastewater to the total number of enterprises. Draft plans for the protection and rational use of water resources should be developed and provide for tasks to stop the discharge of polluted wastewater through the introduction of advanced technology, waste utilisation, construction of treatment facilities. This can be traced on the example of the Ak-Buura River, which flows in the territory of Nookat and Kara-Suu districts of Osh province. It is 148 kilometres long and has a catchment area of 2530 square kilometres. The main source of water supply is snow melt and glacial water. The average annual water flow is 21.4 m^3 /sec. On the Ak-Buura River there are the town of Osh, villages: Papan, Osh, Osh. Osh, the villages of Papan, Toloykon, Ozgur, etc. are located on the Ak-Buura River. It is widely used for irrigation and water supply. The Papan reservoir was built in the middle reaches. Annual runoff is 675 million cubic metres.

Creation of the Papan reservoir on the Ak-Buura River resulted in changes in the hydrochemical regime and drinking water quality. A design error was neglecting the expected mineralisation of water in it as a result of the formation of an artificial lake.

Under conditions of Osh oblast, typical for temperate and warm climate, water "blooms" with blue-green algae, which means that toxic substances are formed in it, water quality deteriorates, extraneous odours and unpleasant taste appear. The water becomes unfit for drinking to some extent. During the periods when a large mass of micro- and macro vegetation dies off, water quality deteriorates even more sharply, dissolved oxygen content decreases, and unpleasant odours appear.

To maintain water quality in the reservoir, the reservoir bed should be thoroughly cleared of vegetation and contaminants annually, after the growing season before filling.

As a result of the formation of the Papan reservoir, the groundwater level has risen in the neighbouring settlements of Kara-Suu and Aravan districts and the western part of Osh city, which floods houses and destroys various buildings.

In addition, the Papan reservoir is located in the 8-9 point zone. This strongly dictates the necessity to keep its dam in serviceable condition, to establish ecological monitoring, systematic forecasting, occurrence and development of ecological problems, possibility of new forms and foci of dangerous geological processes and phenomena, their rates and final results: justification and selection of medical, engineering and economic measures to prevent processes creating ecological problems to combat their average impact, obtaining calculation indicators for their design.

Water supply and sewerage have a significant impact on the environment. Properly addressing these issues has a positive impact on the city's economy and the possibility of its further expansion.

In the cities and working settlements of the region, due to the lack of municipal water pipelines, the population is forced to use technical water pipelines that do not have treatment facilities and water disinfection installations. Due to the lack of water pipelines, 70 per cent of the population of Osh oblast uses directly from rivers, aryks and even various technical water reservoirs for drinking purposes.

Water supply and sewerage networks require large amounts of money to operate efficiently. Water treatment requires aluminium sulphide, liquid glass, soda ash, and liquid chlorine. In addition to purchasing reagents for water treatment, considerable funds should be allocated for repairing water supply networks and laying new main water supply networks. The existing water supply and sewerage networks are worn out.

At present, the problem of water supply to Kyzi-Kii is acute. The water pipeline, which has been under construction for several years, already requires reconstruction. It is necessary to complete anti-fall measures. As it is known, mudflows pose a danger to the life of the town.

The general environmental situation requires construction of a reservoir and treatment facilities in Uzgen. Approximately the same environmental situation in Lyailyak, Batken and Kara-Kulja districts. The lack of treatment facilities and sewerage in these districts threatens the outbreak of new diseases.

In the town of Kara-Suu, treatment facilities and a sewage system with a capacity of 1500 m^3 of sewage per day have been commissioned. These facilities, before starting to work, are already leaking. Hydraulic tests have not been carried out.

As noted above, there have been global weather changes recently. This is connected, in our opinion, with appearance of Kampyr-Ravat (Andijan), Papan, Bazar-Korgon, Tort-Kul, Toktogul (Kairak-Kum in Leninabad province of Tajikistan) and other reservoirs in Fergana valley in the region for the last 20-30 years.

The problem of studying climate formation and the causes of climate change requires consideration of various geological processes.

There are 11 meteorological stations in Osh oblast. In the valley zone, apart from Osh, there are (listed from west to east) in Isfana, Batken, Khaidarken, Naukat, Kara-Suu, Uzgen. There are also high-altitude stations located in Kichik-Alai, Daraut-Kurgan, Sary-Tash, Kyzyl-Jar (this last one is in Kara-Kuldjin district, almost on the southern border).

There are 11 more gauging stations on the rivers Isfayram-Sai, Shakhimardan, Kyrgyz-Ata, Aravan-Sai, Zhazy (Iasi), Zerger, Donuz-Too, Kolduk (the last three rivers are tributaries of Iasi), as well as on the Kara-Kulja, Tara, Kurshab.

There are four agro-posts in the region: in Markaz (Kadamjai district), Karavan (Naukat district), Aravan and Gulcha.

The existing network of meteorological stations, hydro and agro-posts is insufficient.

This list does not include such services as snow avalanche stations, which are so necessary in high-mountainous areas. As research shows, at least two are needed: in Kara-Kulja and Alaikuu.

The reliability of meteorologists' forecasts is currently reduced by the fact that the forfost of this service, the high-mountain station at Chaar-Tash, has ceased to function. It worked until 10 October 1993. As a result of the station's shutdown, the data required by the whole republic are no longer received.

This little analysis argues that lack of funds, economic crisis has led to the collapse of the most important environmental services in the country.

Chaar-Tash station is located at an altitude of 2,748 metres above sea level. The annual precipitation here is 1080 mm per year. For comparison, let us cite the Osh minimum - 353 mm, about three times less. Many watercourses originate in this high-mountainous area on the slopes of the Fergana Range. Many spring cataclysms and natural disasters originate here. For example, in March 1994, landslides occurred in Tosoi and Komsomol villages of Uzgen district. Earlier, warnings to the population about possible disasters were received from Chaar-Tash. But the economic crisis put this environmental problem before the government.

Analysis of preparation and implementation of measures on prevention and elimination of consequences of natural disasters shows that in the conditions of ongoing economic reforms in the Republic, the material and technical base of civil defence has been sharply reduced and, accordingly, its ability to ensure day-to-day readiness to perform suddenly arising tasks.

Joint-stock companies, peasant associations and farms are practically not engaged in solving the tasks to ensure the protection of their workers and employees in case of emergencies. Training of the population on the Civil Defence programme has been stopped.

A complex and problematic issue is the construction of protective structures - shelters, anti-radiation shelters, as well as their maintenance in readiness to receive the sheltered population.

The storage terms of special Civil Defence equipment of the mobilisation reserve stored in the standard warehouses of Osh, Uzgen, Nookat and Batken districts are expiring. Work on timely control of the quality of storage and the renewal of gas masks, children's protective chambers and radiation and chemical reconnaissance devices is not carried out because of a lack of funds. The problem of creating stocks of food and essential goods for the population in case of emergencies in peacetime and wartime deserves special attention.

No less important is the problem of warning the population about the threat of possible situations. However, in connection with the programme on privatisation and denationalisation of state and communal property, 57 out of 103 radio broadcasting stations in Osh oblast were transferred to small enterprises and communal farms. Almost all of these nodes are out of operation, and there is a sharp reduction of radio sets at enterprises and in the homes of the population. There are no stocks of street loudspeakers at the communication centres. All this has led to a sharp reduction in the main method of notifying the population about emergency situations.

Along with very severe problems associated with landslide movement, there is a real danger associated with a source of increased radioactive hazard. This is the tailing dump (technical term) of former mines in Maili-Suu town, Sumsar and Shakaftar settlements in Ala-Buka district, Aidarken and Kadamjai - in Kadamjai district. If even a small landslide occurs here, the consequences may affect the gene pool of the people not only in the Fergana Valley, but also in the whole of Central Asia. The issue of ecological situation requires a big economic solution. According to our research, the environmental consequences of different mining operations in the south of Kyrgyzstan really exist: in Mailuu-Suu, in the virgin Tuya-Moyun of the Aravan district, and at dozens of other ore deposits of various radioactive elements. Here it is necessary to take into account that natural sources of rather strong radioactivity exist in any area and are connected with manifestations of ancient volcanism. New problems of additional inversion have appeared.

The dams and other structures protecting the tailings from the former concentrators must be maintained in good order, not only because of environmental hazards but also for economic reasons. Integrated processing of raw materials has not been fully established in the past. The possibilities of recycling these supposedly "waste" rocks to make them even safer should be explored.

Near such places there are often livestock breeding facilities, cattle grazing, buildings and structures for various purposes. However, all information about natural sources of radioactivity, not to mention the places where radioactive ores are mined, is strictly secret, and it is necessary to find solutions to declassify it: it concerns the health of the population and the economic well-being of the country. In connection with the nuclear weapons tests at the Lobnor test site of the People's Republic of China, it is imperative to study the radiation background in the region with this factor in mind and to take practical measures.

Recommendations for environmental policy reform in the South of Kyrgyzstan include, in addition to general proposals, specific measures such as coordination with neighbouring republics, management of nature reserves located on the border; measures for neutralisation and disposal of hazardous waste, especially uranium deposits; provision of drinking water; improvement of technologies and soil conservation in the district, etc.

3. Improvement of the economic and legal mechanism and management activities in the use of natural resources

The development and improvement of the economic and legal mechanism should be considered as directions for improving the management of natural resources. The main idea here is the legal regulation of relations in the use of natural resources through administrative and legal means with economic incentives. In practice, as before and now, one of the key problems is the confrontation between ecology and economy in the use of natural resources. So far, it has been tried to solve it mainly through the application of administrative-legal methods of influence on the basis of prohibitions, restrictions, measures of criminal and administrative punishment.

If the method of administrative and legal influence is based on the relations of power and subordination, the economic mechanism is based on the material interest of the executor in achieving the real goal. The economic mechanism includes both permanent institutions, without which it cannot function normally, and new ones that have emerged on the basis of transitional relations to market relations.

The concept of economic mechanism in the use of natural resources and environmental protection is a new phenomenon in practice.

The acuteness of this problem can be seen, for example, in the issues of charging for resource use, including fees for collecting fresh water and fees for discharging polluted effluents into water bodies.

In the post-Soviet economic literature, the principle of water charging is largely unobjectionable; however, the methodological approaches to determining the payment for water and wastewater discharge are diverse.

There are two main approaches to the definition of fresh water. According to the first approach, the assessment of natural resources is linked to the costs of their development and reproduction. Since labour is used in the use of water resources, water has a social value. Determination of water price is carried out similarly to the calculation of the cost of water and profit is found on the basis of the average national economy profitability rate. The calculation of the price includes the costs of water supply and water disposal (costs of water intake, transport, local treatment and waste water disposal), for the study and protection of water resources, for their regulation, etc.

Practical conditions of this group of economists are aimed not at optimisation of water resources use, but at "legalisation" of water sector costs.

According to the second point of view, water, like any natural resource, should be valued taking into account its limited and multivariable use. The price of water should reflect the added economic benefit arising from its use. Hence, water valuation arises as a direct function of differential rent, which shows the amount of economic gain for the national economy.

The fee for wastewater discharge is determined either on the basis of the price of fresh water and the volume of water required to dilute the effluent to its natural state, or it is linked to the amount of economic damage from water pollution. The first approach is increasingly difficult to realise practically, as the condition of water resources already in natural water bodies often does not meet the requirement.

The concept where the fee is considered as a control parameter that stimulates enterprises to choose the most efficient from the national economic point of view by obtaining the required environmental quality is gaining more and more supporters. However, it should be remembered that the introduction of water use fees is unlikely to solve environmental problems unless it is supported by radical changes in the economic mechanism as a whole and the current profit distribution system, in particular.

The current fee system in the water sector is insufficient for optimising water use because it does not take into account the economic assessment of water as a natural resource.

In our opinion, it is necessary to establish a tariff separately for water: large rivers; in canals; water supplied by the water supply system to the population, water supplied by the water supply system to industry (within the norm and above the norm); recycled water and water in irrigation systems.

Along with water charges, it is necessary to introduce water pollution charges. The ways of formation of pollution charges proposed in the literature are very diverse, the charges are linked to the damage caused by wastewater discharge, to the volume of excessive discharges, to the cost of wastewater, etc. At present, there is no system of payment for pollution of water bodies in Kyrgyzstan. Enterprises pay only for wastewater discharge into sewerage systems and municipal treatment facilities (treatment fee). Legal (criminal, civil, administrative) liability measures for pollution of water bodies and water resources provided for by the legislation open wide opportunities for increasing the efficiency of the whole social production and water sector as one of its branches.

In the development of the economic mechanism for the use of resources, it is important to achieve minimum costs of social labour, rational distribution and use of extracted resources, preservation of the environment at a level that ensures normal conditions for the life of society, now and in the foreseeable future.

In the most general form, the scheme of the most intensive links between individual sectors of the economy and nature management is presented in Table 15.
Scheme of linkages between economy and nature management

Table 15

Economic sectors	Elements of nature	The nature of the links
1	2	3
Agriculture	Earth, soil, atmosphere, climate, plant life	Utilisation for production purposes. Reproduction of soil fertility by applying organic and mineral fertilisers
Mining industry	Subsoil, land, landscape	Land subsoil use for industrial purposes, landscape change in open-pit mining operations
Energy industry	Atmosphere, waters, earth	Use of atmospheric air, water, emission of waste into the atmosphere, artificial water bodies
Chemical and petrochemical industry	Waters, atmosphere	Use of water resources, atmospheric air, waste emission into the atmosphere
Forestry, timber, wood processing industry	Forest, water, soil	Reproduction of forests, use of forests as raw materials, changes in flora and soils
Fish industry	Waters, wildlife	Use of fish resources, water bodies, reproduction of fish stocks. Changes in fauna
Land and air transport	Atmosphere, earth	Use as area, atmospheric air, emission of waste into the atmosphere

In terms of the intensity of interrelationships, or more precisely, in terms of dependence on nature, agriculture ranks first. Nature, in fact, is the main factor and the most important condition of agricultural production. The main means of production in agriculture is land. If the process of production of industry, regardless of soil fertility, the quality of natural vegetation and many other properties inherent in the land, then in agriculture the process of labour and the creation of products is connected precisely with these properties of land, with its qualitative state, with the nature of its use. What is the quality of the land, what is the level of its utilisation, the yield and volume of agricultural products largely depend on it.

Extractive industry ranks second in intensity in relation to the environmental economy and its dependence on it. It also, like agriculture, deals with the substance of nature, and with the substance of its primordial world. The object of its activity is mineral wealth. The interaction of the extractive industry with the earth's subsoil and surface has not only direct but also inverse relations.

At present, about 30 types of basic minerals out of the 200 most known are used in Kyrgyzstan.

The use of water resources on a large scale in the production processes of chemical enterprises causes discharges, waste water and water polluted with various chemical substances into the water environment (dyeing shops of the Osh silk plant and the production cotton plant).

In the area of environmental management, the most common measures so far remain measures of administrative responsibility combined with some economic measures (fines, warnings, bans, temporary closure of enterprises or production, reimbursement of costs for the use of resources and other types of economic sanctions).

The economy of the Kyrgyz Republic is currently experiencing a deep crisis. With the acquisition of political sovereignty, extensive internal and external economic ties between the republics were severed.

Due to the lack of raw materials and component elements of production, the country's industry has become an unviable branch of the national economy. A course on denationalisation of industrial facilities and introduction of private property was taken. An anti-monopoly policy is being pursued against large industrial enterprises.

Construction organisations have been stopped due to the inability of economic entities to produce construction materials based on local raw materials. Capital investments have no source of formation. The existing production facilities and contracting organisations have actually turned out to be unnecessary, and the volumes of orders of new economic entities are infinitesimal, insignificant, not static, so it is possible to ignore them.

Along with the economic mechanism of natural resources management, the legal support of effective nature management and, first of all, the organisational structure of environmental management are of great importance. In the republic, natural resource management issues are closely intertwined and sometimes form an integral part of nature protection problems. Since independence, the country has taken certain steps to bring the organisational structure of environmental management in line with the needs of the new state. The President's Parliament, the Government and local Administrations are responsible for nature protection.

The Parliament (Jogorku Kenesh) is responsible for 1) defining environmental policy, 2) developing and adopting laws and resolutions on nature protection issues, 3) approving the Government's proposals on the limits of fees for the use of natural resources and taxes. During the first year of independence, the Parliament had the Jogorku Kenesh Commission on Ecology, which had three permanent members. Due to financial difficulties, it was merged with the Commission on Agrarian Issues, Food and Ecology, which was responsible for coordinating all changes in legislation on environmental issues and submitting issues on nature protection to Parliament.

However, due to the fact that agriculture is a key sector of the Kyrgyz economy, the problems of the agricultural sector dominated the work of this commission, despite the fact that the subcommission on environmental issues was very active[1] .

From the same document it is possible to find out the responsibilities of each management structure. Thus, the President of the Republic also deals with environmental issues. The President has the following competences:

1) establish a regime of special nature use, 2) declare and establish the boundaries and regime of zones of ecological emergency or ecological disaster, 3) approve the procedure for the formation and use of nature protection funds. The President also signs all laws adopted by the Jogorku Kenesh, including laws on nature protection. He conducts international negotiations, signs interstate treaties, and submits them to the Jogorku Kenesh for ratification.

As for the Government of the Republic, it and its numerous ministries and departments are directly involved in the issues of nature management. Today there is not a single ministry and agency that has not directly or indirectly dealt with the issues of nature use and environmental protection in its activities. Nevertheless, the main structural unit of the Government responsible for environmental policy is the Ministry of Environmental Protection, originally established as Goskompriroda back in 1988 as part of the former Union Government's attempt to strengthen environmental organisations by replacing the Government Commission for Nature Protection and giving them the status of a ministry. Pursuant to the Law on Nature Protection, the Government of the Kyrgyz Republic issued Resolution No. 71 of 16 February 1994, which defines the status of this agency within the Government and its functions. In the approved regulation, it is assigned, inter alia, the following responsibilities:

1) implementation of integrated management of environmental protection activities in all sectors of the national economy, 2) organisation of development and approval of environmental standards, norms, rules, regulations and instructions on nature protection, 3) development and submission to the Government of the Republic of proposals ensuring active impact on investment processes related to nature management. According to the regulation, the organisational activities of the Ministry are related to

1) creation of an environmental information data bank on the state, use and protection of natural resources, 2) accounting of emissions, discharges of pollutants, harmful effects on the environment, 3) establishment of limits and standards of nature use, issuance of permits for nature use, 4) conducting state environmental expertise, 5) organisation of activities of nature reserves, natural national parks.

Local state administrations also play a significant role in natural resource management. Three territorial departments of environmental protection are responsible for environmental management in six oblasts.

Thus, many management bodies, both at the republican and regional levels, deal with the problems of nature management to a greater or lesser extent. The task is to orient all management bodies and economic units on qualitative fulfilment of legislative provisions in the field of nature use on the one hand, on the other hand, to ensure flexible response of these same bodies to market relations.

Another problem in environmental management is the imbalance of authority and responsibility in decision-making on the part of various management bodies. The management system is still dominated mainly by administrative methods of management, which are essentially command-and-control. Nowadays, the rights of economic entities have been significantly expanded, and, consequently, the functions of various management bodies in the field of nature management have also become diverse.

Among the management problems, mitigation of the impacts of natural disasters within the framework of accessibility occupies a very important place. Thus, the entire territory of the republic is constantly exposed to the active impact of natural processes leading to disasters and human losses. Out of 70 types of cataclysms widespread in the world, 20 are manifested in Kyrgyzstan.

More than 3000 earthquakes are registered in Kyrgyzstan every year, of which on average ten strong earthquakes occur, and every 5-10 years there are devastating seismic disasters; about 38% of settlements are located in the zones of possible earthquake sources with intensity of 9 and more points. For example, over the last 100 years, 100 strong earthquakes have occurred in the city of Osh.

Another element is mudflows and floods. Landslides and landslides occur frequently in particular in the South of Kyrgyzstan, where in some areas the affected area reaches 30-40 landslides per 1 square kilometre.

Frosts and snowfalls fall on vast areas of agricultural lands 2-3 times a year. From 1950 to 2014, 72254 avalanches were registered, every fiftieth one caused damage to the national economy and killed many people. This means that research work to identify the most likely areas of landslides, debris flows, etc., as well as early detection of avalanches and avalanches. and also early warning of the population and economic entities should be elevated to the rank of the most important tasks in nature management.

Meanwhile, environmental management problems are often reduced to a narrow interpretation. In many cases, environmental management problems are understood as protection of water, air, soil from pollution, etc. Undoubtedly, protection of water, air, soil from pollution and maintenance of ecological balance, and simply nature conservation are all components of environmental management. However, it is not limited to it alone. Environmental management is the management of its resources with the aim of ensuring their rational use while maintaining ecological balance. As such, the environment acts as a peculiar branch of the national economy or as a meta-infrastructure of social production.

Since the problem of environmental management is closely related to the management of environmental protection, two sides of these interrelated processes should be distinguished: management through the organisation of economic activity and management directly by the object and the environment and nature use. In the first case it is closely connected with the general management of the economy, in the second case it acquires the character of purposeful impact of society on certain objects of the environment and nature use. The limits of such impact are determined, on the one hand, by the development of science and technology, cognition of natural laws of nature, cause-and-effect relations in the environment, and, on the other hand, by the requirements arising from the need to preserve and improve the environment and nature management.

Accordingly, it is necessary to improve not only organisational forms of nature management, but the whole complex of its legislative, regulatory, informational, economic and educational aspects.

The specificity of nature management in Southern Kyrgyzstan is determined by relatively high population density, conditions of industrial and agricultural development, and diverse natural conditions. Environmental protection activities consist of measures to protect air, water, subsoil, lands, forests, flora and fauna, and natural monuments.

The Osh-Jalal-Abad Department of Environmental Protection exercises direct control over the state of the environment.

Atmospheric air protection should be ensured by installation of treatment facilities at industrial enterprises, boiler houses and other sources of air pollution by vehicles. Development of centralised heat supply in cities and large settlements, expansion of green areas, installation of diagnostic equipment in motor vehicles and other measures contribute to improvement of the atmospheric air condition.

Water protection includes measures for wastewater treatment and use for irrigation. Hundreds of treatment facilities operate in the region, providing treatment of 94.9 million m^3 of wastewater (of which 62 million are normatively clean and the rest are normatively treated).

Subsoil protection should be ensured by the introduction of scientifically sound technologies that promote more complete extraction and integrated utilisation of mineral resources at developing mineral deposits.

Brick factories in Sulukt, Kyzyl-Kiye, Osh, Kochkor-Ate and Myrza-Ake have been built on the basis of using roof rocks. However, not all of them are functioning normally.

Protection of land resources includes measures on land reclamation, watering, prevention of erosion processes, introduction of new soil-protective methods of soil cultivation, creation of field-protective forest strips, ravine and gully green plantations.

The protection of the region's forests is especially important due to their uniqueness and the extremely important role of water protection, soil protection and erosion control in the foothills and mountain slopes, where the main massifs of wild fruit and juniper forests are concentrated. Much work on the protection and restoration of forests is carried out by regional forestry enterprises and the Southern Kyrgyz Department of Nut and Fruiting Forests.

Protection of flora and fauna includes measures for the conservation and reproduction of plants and animals. The most effective is the creation of a network of specially protected areas, dispersal, creation of tree nurseries, etc. Southern Kyrgyzstan is inhabited by species included in the "Red Book" of the Republic: 52 species of animals and fish, 43 species of plants.

Specially protected areas are created for the purpose of preserving natural complexes or objects of economic, scientific, aesthetic, cultural and educational significance. On the territory of the region they include: Sary-Chelek State Biosphere Reserve, Osh forestry reserve, 5 ha reserve in Alai district, 30 ha Ak-Bura reserve in Nookat district; geological reserves and natural monuments: Sulaiman-Too, Abshir-Ata waterfall, Chil-Ustun cave and others.

The legal framework of environmental management in environmental protection is extensive and diverse. The Constitution of the Republic defines the rights and obligations of citizens, officials, authorities and management bodies in the field of nature use.

In management activities, environmental certification, rationing and limitation, payment for use of natural resources and pollution, environmental control, release of finished products and waste disposal are important.

At the same time, environmental passportisation and environmental control do not require large investments for their implementation. Timely and quality implementation of these works would contribute to: drawing up a list of enterprises most responsible for pollution and defining priorities for future regulatory activities in the industrial sector; conducting the so-called environmental audit of enterprises and elaboration of an environmental action plan to be introduced into the business plan of the enterprise; disposal of harmful industrial waste, etc.

There are a number of other channels for improving management performance. For example, preventive maintenance, which is always cheaper and more effective than correcting the consequences of an event that has occurred. The introduction of improved and more economical technologies and waste treatment, etc., is very useful in this regard.

Meanwhile, the State Sanitary Supervision of the Republic is concerned about the continuing deterioration of preventive and sanitary works, which leads to a decline in the level of operation of sewerage and water supply networks. In 1994, 145 out of 1022 water pipelines were in unsatisfactory condition, including 9 communal and 136 departmental ones. There were no sanitary protection zones for water sources and water intake structures on 90 water pipelines and no disinfection facilities on 63 water pipelines. Most of the existing water pipelines were built 40 and more years ago and have not been overhauled, about 40% of the networks are worn out and should be replaced and reconstructed.[1]

These are some measures to improve the economic and legal mechanism and intensify management activities in the field of natural resource use.

We propose the following scheme of environmental management bodies of the Kyrgyz Republic.

Conclusion

The paper studies the problems of nature management, which involves science-based regulation of society's impact on nature, preservation, restoration and improvement of favourable natural conditions of human life, development of production and its culture. This issue in the conditions of transition to market relations should be among the most important problems of socio-economic development, and requires the development and implementation of measures aimed at the organisation of production, achieving the greatest results at the lowest cost of funds and natural resources.

Based on the results of the study, the following conclusions and suggestions can be made.

1. The problems of natural resources management are currently acquiring special relevance. This is due to the fact that natural resources create, on the one hand, the very possibility of human activity and, on the other hand, have a profound impact on all aspects of social life, economic and technical progress.

The peculiarity of recent decades is that the negative consequences of progress in nature management are increasing. In the coming period, in order to ensure economic growth, it is necessary to concentrate resources for the implementation of the most important national programmes, to find ways of their rational use, to strengthen the regime of economy, to introduce market mechanisms for the management of national natural resources. This is necessary to solve social problems, improve the living standards of the population and, above all, to protect the health of the individual.

2. Nature management occupies an important place in the system of management of national economy. It has been established that the basis of natural resources management is their rational use and the complex of society's relations to

nature, including the preservation, restoration and multiplication of renewable natural resources, the creation of environmental protection measures of economic and legal mechanism, coordination of relations between environmental protection bodies, numerous industrial, agricultural and other enterprises and the state. Improvement of natural resources management implies the development and implementation of aggregate economic, managerial and legislative measures aimed at more appropriate organisation of production processes, rational and integrated use of resources, including in the regional context.

3. The research clarified and systematised the main criteria and principles of natural resource management. The main criterion of natural resource management is the protection of human health, which should be enshrined in legislative acts concerning the use of natural resources and environmental protection.

The next criterion of resource management is socio-economic growth and development. This criterion is important for Kyrgyzstan, given the need to improve the living standards of the population and reduce poverty.

Another important criterion for natural resource management is the impact of management decisions on the introduction of market relations. In the transition period, the evaluation of management decisions on this indicator should be intensified.

Important criteria for resource management include assessing the efficiency of replacing primary resources with waste, as well as developing the capacity of human resources and increasing their civic and administrative responsibility.

4. To improve manageability, natural resources need to be assessed. It is necessary to establish resource accounting, environmental information and environmental management mechanisms.

With regard to the South of the Republic, it should be said that most of the territory is occupied by mountains, while the plains account for only 7% of the total area and are characterised by dry and hot climate, high dustiness of the air. The average population density is 28 people per square kilometre. However, the region is very unevenly populated. Mountains, forests, reservoirs, mountain rivers and mountain lakes, minerals and others, form the basis of natural resources.

The analysis has shown that many resources involved in economic activities are used to the detriment of the environment, and production activities are often accompanied by violations of environmental requirements. For example, the production of antimony and its compounds, as well as mercury, occurs with severe air pollution. Coal consumption on a large scale also results in increased dust and gas emissions into the atmosphere.

The assessment has shown that the territory of the South is rich in deposits of mineral resources. However, they are currently insufficiently studied and explored. Therefore, it is necessary to carry out a comprehensive exploration of the deposits.

5. In the system of controllability of nature management, an important place is occupied by the use of land, which is the initial material basis for the well-being of society, the spatial basis for the location of productive forces, the basis for the normal course of reproduction processes of all factors of economic growth.

The study has shown that land resources are used far irrationally and in many ways inefficiently. Despite calls to intensify the movement in favour of environmental protection measures, in the post-Soviet period the environmental situation has generally deteriorated. Due to the lack of funds, many environmental protection measures to improve the use of land resources are not implemented, the discharge of polluted water is increasing from year to year, the amount of waste is increasing, and measures to improve soil fertility, the fight against erosion and spoilage of land cover are carried out unsatisfactorily. All this affects not only the state of agrarian economy, but also the indicators of the whole national economy.

There are big problems in the use of water resources, especially in its consumption as drinking water. A number of cities and settlements use contaminated water as drinking water. The situation is aggravated by the fact that the effectiveness of biological means for wastewater neutralisation is not systematically assessed in the region. The utilisation of biological resources is also unfavourable due to habitat degradation. All this requires the adoption of rational utilisation of natural resources.

6. State regulation is important in the management of national natural resources. The economic and regulatory functions of the state are diverse. They in the use of natural resources are mainly related to the provision of legal framework, environmental protection and conservation, foreign economic activity.

State regulation of resource use is carried out in various forms through distribution, development of normative documents and legal provisions, establishment of economic sanctions, adoption of managerial decisions, control, etc. However, studies show that there are many problems of legislative, normative, organisational and managerial nature in the republic. From year to year, for example, monitoring and control over the use of resources are deteriorating due to a sharp decline in the sources of financing. There is an urgent need to adopt laws on nature, water, land, etc.

7. Among the most important ways to improve the management of national natural resources are measures to raise living standards and improve human health. At present, according to official statistics, 91.8 per cent of the population of the Republic have incomes below the minimum consumption, and therefore below the poverty line. At the same time, over 70 per cent of the population has an income below 175 soms per month, making up the stratum of the extremely poor.

The main ways to improve the living standards and incomes of the population related to the use of natural resources are:

- Improving the use of land resources by adhering to agrochemical practices, applying sufficient doses of fertilisers, adopting advanced technology, soil conservation, improving irrigation practices, increasing soil fertility, etc;

- successful implementation of land and agrarian reform in the republic and in the regions;
- development of the basic principles of the market mechanism of economic management and their implementation;
- Improved utilisation of forest resources;
- development of tourism, especially ecological tourism;
- A strong industrial policy that provides for the development and integrated utilisation of natural resources.

As ways to improve the health of the population, the paper substantiates measures on environmental health care, which in a market economy can be developed both through state support and financing and private medical practice. They should provide for the maximum elimination of the influence of harmful natural substances, as well as water containing bacteria and viruses. Timely collection and processing of information on the morbidity of the population in order to take specific measures also plays a positive role in protecting human health.

8. Improving the management of natural resources includes measures to improve environmental protection, the essence of which is to ensure, through scientifically-based regulation of society's impact on nature, the preservation, restoration and, where necessary, improvement of favourable conditions for human life, production and cultural development. Problems of environmental protection are largely related to the work of industrial enterprises and farms in the agricultural sector. One of the ways to positively influence the work of these enterprises is the strict implementation of legislative provisions in the field of environmental protection. In the coming period, it is necessary to achieve rational use of water resources and preservation of cleanliness of natural water bodies, as well as payment for the use of natural resources.

The task of preventing and eliminating the consequences of natural disasters: gravitational shifts, mudflows, earthquakes is quite peculiar. This requires the adoption of complex engineering and economic measures. The problem of environmental propaganda and education is also among the measures to improve the manageability of nature management.

One of the practical steps of environmental protection is the implementation of measures of the National Plan on this problem (NPEP), adopted by the Government of the Republic in 1994 with the active assistance of the World Bank. At the same time, the analysis has shown that the deadline for implementation of the NEPP ends in 1997, but work has not yet begun on many aspects of the plan.

9. The work of environmental management bodies should be significantly

improved. The task is to orient all authorities and economic entities to the qualitative implementation of legislative provisions in the field of nature management, on the one hand, and to ensure flexible response of these same bodies to market relations, on the other hand.

Significant work remains to be done in the creation of an economic mechanism for natural resource management. If the method of administrative and legal influence is based on the relations of power and subordination, the economic mechanism is based on the material interest of the executor in achieving the real goal. Recommendations in this case concern the issues of creating an environmental fund, payment for resources, fines and sanctions for environmental offences, taxation and financing.

The specificity of nature management in Southern Kyrgyzstan is determined by relatively high population density, conditions of industrial and agricultural development, and diversity of natural conditions. In this regard, environmental passportisation and environmental control are of great importance. The importance of preventive work in environmental protection activities, protection of atmospheric air, water, subsoil, lands, forests, fauna and flora is extremely high. The analysis shows that the structure of environmental protection activities in the Kyrgyz Republic is rather complex, diverse, multidepartmental and not of a pronounced departmental or territorial nature, as well as it does not meet the requirements of the new economic mechanism in a market economy in many respects.

The current Ministry of Environmental Protection deals with a very narrow range of issues, partly such as imposing fines on those responsible for environmental pollution. The State Agency for Geology and Mineral Resources controls the subsoil itself, land - the Land Inspection, water - the Sanitary and Epidemiological Inspection. There are also various kinds of fish and hunting inspections, meteorological, seismic stations and others. Animal and plant protection is also not specific. Measures against geological hazards are handled by the Ministry of Emergency Situations without a counteracting mechanism that meets the requirements of the law of unity of opposites.

As a result, important environmental problems in Kyrgyzstan remain out of sight of the state and its government. In this regard, it is necessary to develop a National Programme of Natural Resources Management with appropriate legislative acts providing for legal and economic mechanisms and to establish a Ministry of Ecology, which could coordinate the work on environmental protection and exercise control over the work on the rational use of natural resources.

LIST OF REFERENCES

1. Aganbegyan A.G. Scientific and technological progress and acceleration of socio-economic development. Moscow, Economics, 1985.
2. Arbatov A.A. In defence of natural resources of the planet. M., Znanie, 1985.

3. Abramov S.P., Zaleskiy F.V. Engineering surveys in construction. M., Stroyizdat, 1982.
4. Ananyev V.P., Korobkin V.I. Engineering Geology. M., 1973.
5. Aleksandrovskaya Z.I. Blagodrustvo Gorodstvo. M., Stroyizdat, 1984.
6. Aral: from words to deeds. Materials of the Nukus Declaration on the Aral Sea Basin Development, 18-20 September 1985.
7. New Agrarian Policy. Speech of the President of the Kyrgyz Republic at the solemn meeting dedicated to the Day of Workers of Agriculture and Food. Slovo Kyrgyzstana, 9 November, 1995.
Akaev A.A. On the strategy of socio-economic development of the Kyrgyz Republic and urgent actions. Bishkek, 1993.
8. Assakunov B.T. Decorative cements and concretes from local raw materials. Bishkek, Kyrgyz Scientific Research Institute, 1991.
9. Botolov NA, Mitroshin K.P., Shaposhnikov A.K. Protection and utilisation of biological resources in the USSR. M., Znanie, 1970.
10. Baibosunov A. Pre-scientific representation of the Kyrgyz about nature. Under general editorship of Doctor of Philosophical Sciences Narynbaev A.I. Frunze, Mektep, 1990
11. Belov S.V., Barbinov F.A.. Koziakov A.F. et al. Environmental protection. Moscow, Higher School, 1983.
12. Berdyaev N. Entry of the machine "Nature and Man", 1989, No. 8.
13. Balatsky O.F., Ermolenko B.A., Zaitsev V.A. et al. Waste-free production: economics, technology, management (Results of science and technology).
T. 17, M., VINITI, 1987.
14. Biosphere, its present, past and future. M., Prosveshchenie, 1976.
15. Belyaikin K.M. Cheap energy, by-products, clean air. "Chemistry and Life", No. 9, 1986.
16. Brylov S.A., Strodki K. Environmental protection. Textbook for students of mining and geological specialities of universities. Moscow, Higher School, 1985.
17. Bystrakov Y.I., Kolosov A.V. Economics and ecology. M., Agropromiz-dats, 1988.
18. Bekker A.A., Agaev T.B. Protection and control of pollution of natural environment. L., Gidrometeoizdat, 1989.
19. Besserer D. On the inefficiency of the tourism support system in Kyrgyzstan. Slovo Kyrgyzstana, 5 October 1995.
20. Vladimirov A.M., Lyakhin Y.I., Matveev L.T., Orlov V.T. Environmental protection. Textbook for students of higher educational establishments. L., Gidrometizdat, 1991.
21. Gilbert White. Geography, resources and the environment. Selected articles.

Moscow, Progress, 1980.
22. Golubev I.R., Novikov Y.V. Environment and transport. Moscow, Transport, 1987.
23. Gromov B.V., Zaitsev V.A. et al. Waste-free industrial production. Basic principles of waste-free productions. "The results of science and technology". T.9.M., 1981.
24. Gladkiy I.I. Human factor of development. M., Economics, 1986.
25. Grigoriev N. What kind of market do we need? Slovo Kyrgyzstana, 2 November 1995.
26. GOST 17.5.1.02-85. Nature protection. Lands. Classification of disturbed lands for reclamation. M., Standards Publishing House, 1985.
27. GOST 17.4.3.02-85. Nature protection. Soils, Requirements for the protection of the natural soil layer during earthworks. M., Standards Publishing House, 1985.
28. GOST 17.2.1.-04-77. Nature protection. Atmosphere. Sources and meteorological factors of pollution. Industrial emissions. M., Standards Publishing House, 1984.
29. GOST 17.2.4.05-83. Atmosphere. Nravimetric method for determination of suspended dust particles. M., Standards Publishing House, 1984.
30. GOST 171.3.11-84. Nature protection. Hydrosphere. General requirements for protection of surface and ground waters from pollution by mineral fertilisers. M., Standards Publishing House, 1984.
31. GOST 17.1.3.10-83. Hydrosphere. General requirements for protection of surface and underground waters from pollution by oil and oil products during pipeline transportation. M., Standards Publishing House, 1985.
32. GOST 17.5.1.01-83. Nature protection. Land reclamation. Terms and definitions. M., Standards Publishing House, 1984.
33. GOST 17.6.01-83. Nature protection. Protection and defence of forests. Terms and definitions. M., Standards Publishing House, 1984.
34. Davydov I.A. Trump card of our economy. Slovo Kyrgyzstana, 31 August 1995.
35. Dolgorev A.V. Secondary raw material resources in the production of building materials. M., Stroyizdat, 1990.
36. Earth, ecology, new stage of perestroika and literature. Materials of the Plenum of the Board of the Union of Writers of the USSR 17-19 January 1989. Literaturnaya gazeta, 25 January 1989.
37. Legislation on radical restructuring of economic management. Moscow, Yuridicheskaya Literatura, 1988.
38. Ibraimov S.I. Osh oblast of Kyrgyzstan. Frunze, Kyrgyzstan, 1979.
39. Kafarov V.V. Principles of creation of waste-free chemical productions. M., 1982.
40. Kirpatovsky Ch.P. Nature Protection. Reference book for workers of oil refining

and petrochemical industry. Moscow, Khimiya, 1980.
41. Kogan E.I., Haikin V.A. Labour protection at the enterprises of automobile transport. M., Transport, 1984.
42. Comprehensive Programme of Scientific and Technical Progress of the Kyrgyz SSR for 1991-2010. Frunze, 1988.
43. Kondratyev A.I., Mestechina N.I. Labour protection in construction. Moscow, Higher School, 1990.
44. Kuzminov LA., Shvalev L.N. Economic evaluation of labour protection work in construction. M., Stroyizdat, 1982.
45. Lebedinsky Y.P., Sklyansky Y.V., Popov P.I. Resource saving and ecology. Kiev, 1990.
46. Livchak I.F., Voronov Y.V. Environmental protection. M., Stroyizdat, 1988.
47. Liikevich V.K. Dwelling and Climate. M., Stroyizdat, 1980.
48. Lemeshev M. Economy and ecology: the fatal conflict and ways of its resolution. "Voprosy ekonomiki", № 11,1990.
49. Markovich D.J. Social Ecology. A book for teachers. Translation from Serbo-Croatian. Moscow, Prosveshchenie, 1991.
50. Melnikov N.I., Volokov A.I., Korotkova D.A. Pesticides *and* Environment. M., 1977.
51. Mikheev A.V. et al. Nature Protection. Moscow, Prosveshchenie, 1987.
52. Mitikeev K.M. Problems of environmental protection in the light of the decisions of the XXVI Congress of the CPSU (on the materials of the South of Kyrgyzstan). Osh, department. "Znanie", 1981.
53. Mochalov I.P., Rodziller I.D., Zhuk E.G. Purification and disinfection of waste water of small settlements (in conditions of the extreme North). L., Stroyizdat, Leningrad Branch, 1991.
54. Murashko A.I., Stelmashok E.A., Zhalko V.V. et al. Conservation of soils. Minsk, 1989.
55. Musakozhoev Sh. Kyrgyz Republicsynyn ekonomiki. Bishkek, 1997.
56. Myakinnikov V.I., Sultanov A. Ecology and the city: harmony is achievable.
г. Osh, Lenin Way, 6 November 1986.
57. Marx K., Engels F. Op. 23.
58. Nikitin D.P., Novikov Y.V. Environment and Man. Moscow, Higher School, 1986.
59. Novikov Y.V. Nature and Man. M., 1991.
60. On the protection of atmospheric air. Law of the USSR, Collection of Documents of the Party and Government 117-1981. "On the Protection of the Environment". Moscow, Politizdat, 1981.
61. Environmental protection. Edited by Belov S.V. Textbook for universities.

Moscow, Higher School, 1991.
62. From Plan to Market. World Development Report - 1996. International Bank for Reconstruction and Development (World Bank, 1996. 1818 H Street, N.W., Washington, D.C. 20433, U.S.A.
63. Osh region. Encyclopaedia. Editor-in-chief Oruzbayeva B.O. Frunze, 1987.
64. On the protection of the environment. Compiled by. Galeev A.M.. Kurok M.L. Collection of documents of the Party and the Government. Moscow, Izd. of political literature, 1979.
65. On elimination of iodine deficiency disorders among the population of the Kyrgyz Republic. Materials of the Resolution of the Kyrgyz Republic dated 23 September 1995.
66. Pasechnik S.T. Nature, Man, Law. Frunze, Kyrgyzstan, 1986.
67. Petunii V.A. Cities and villages - clean environment. Frunze, Kyrgyzstan, 1990.
68. PertsikN.A. Human environment, foreseeable future. M., Mysl, 1990.
69. Pikhota N.A., Soloshenko I.I., Zakirov Sh.S., Banogin S.N. Problems of rational utilisation of mineral and raw material resources of the South of Kyrgyzstan. Osh, 1994.
70. Plyshevky B.P. Saving means multiplying. M., Znanie, 1986.
71. Confront the elements with the whole world. Materials of the XIV session of the Jogorku Kenesh of the Kyrgyz Republic. Slovo Kyrgyzstana, 13 April 1994.
72. Prodan E.I. Kyrgyzstan - not yet a paradise for tourists? Echo of Osh, 2 September 1995.
73. Pletnev N. How is the health of rivers domestic? Echo of Osh, 1 February 1996.
74. Popov G.H. Problems of the theory of management. M., Economics, 1970.
75. Reimers N.F. The ABC of Nature. Micro-encyclopaedia of the biosphere. M., Znanie, 1980.
76. Remezov N.P., Makarov V.T. Soil science with the basics of agriculture. M., 1963.
77. Rogalsky V.P. Geologists at the crossroads. Echo of Osh, 26 October 1991
78. Rumyantsev A.M. Economy of Nature. Nature, 1977, No. 1.
79. Rybiev I.A. et al. General course of construction materials. M., Higher School, 1987.
80. Stavinsky V., Gorodnyansky I. The riches of the subsoil of Kyrgyzstan. "Goods in stock". "Svobodnye gory", 25 March 1994.
81. Sidorenko R. The Earth does not tolerate emptiness. Slovo Kyrgyzstana, 28 October 1995.
82. Satybekov E. The Element. Case No...., 31 January 1996.
83. Dictionary of Foreign Words. Edited by Lekhin I.V. and Petrov F.N. M., Izd. innatsslovarei, 1949.
84. Savchenko A.P. Secondary resources and waste-free production. M., Znanie,

1986.
85. Smyan N.I., Kudlo K.K., Solovei I.N. Zakazanstvo o zapovedeniya o soils. Minsk, 1984.
86. Noise reduction in buildings and residential areas. Edited by G.L. Osipov. and Yudin EL. M., Stroyizdat, 1987.
87. Sokolov A.A. Water: problems at the turn of the XXI century. L., Gidrometeoizdat, 1986.
88. Reference book on engineering geology. Edited by Churinov. M., 1981.
89. Suerkulov E., Gareeva A. Current state of ecology of Kyrgyzstan. Bishkek, 1997.
90. Tazhibaev A., Abylov S. Tushtuk Kyrgyzstandyn ecologiyasy. 0,1992.
91. Usubaliev T.U. How to divide the souls of peoples by borders? Open Letter to the Presidents of the Uzbek, Kazakh and Kyrgyz Republics. Slovo Kyrgyzstana, 31 October 1995.
92. Ustiyan I. Ecological-economic principle of agrosphere functioning. Journal "Planning economy", No. 12, 1990. - c. 93.
93. Fedorov V.M. Biosphere, Agriculture, Mankind. M., Agropromizdat, 1990
94. Fratanov G.S. Geology and living nature. L., Nedra, 1982.
95. Philosophical Dictionary. Edited by Rosenthal M.M. M., Politizdat, 1975.
96. Philosophical Dictionary. Edited by Frolov I.T., M., Politizdat, 1987.
97. Man and elements. Popular-scientific hydrometeorological collection. Edited by Ugryumov, Kondratov A.M. M., Gidrometeoizdat, 1990.
98. Four-language encyclopaedic dictionary of terms on physical geography. Edited by Spiridonov A.I. M., Social Encyclopaedia, 1980.
99. Economics of Nature Management. Edited by Khachaturov T.S. Textbook for students of economic specialities of universities. Moscow, Moscow State University, 1991.
100. Encyclopaedic Agricultural Dictionary-Reference Book. M., 1959.
101. Ekholm E. Environment and human health. M., Progress, 1980.
102. Environmental Energy and Construction Design. Translated from English by Ivanova G.A. Edited by Bogoslovsky V.N., M., Stroyizdat, 1983.
103. Yusupov T. In order not to stay with "indigenous aryks". Slovo Kyrgyzstana, 16 November 1995.
104. Yagodin G.A., Tretyakova L.G. Chemical technology and environmental protection. M., Znanie, 1984.
105. Yakovlev S.V., Prozorov I.V. Rational use of water resources. Textbook for universities. Moscow, Higher School, 1991.
106. National Human Development Report of the Kyrgyz Republic for 1996. Bishkek, 1977.

MIX
Papier aus verantwortungsvollen Quellen
Paper from responsible sources
FSC® C105338

Printed by Books on Demand GmbH, Norderstedt / Germany